Interdisciplinary Topics in Gerontology

Vol. 40

Series Editor

Tamas Fulop Sherbrooke, Que.

Aging and Health – A Systems Biology Perspective

Volume Editors

Anatoliy I. Yashin Durham, N.C.
S. Michal Jazwinski New Orleans, La.

36 figures, 8 in color, and 9 tables, 2015

Basel · Freiburg · Paris · London · New York · Chennai · New Delhi ·
Bangkok · Beijing · Shanghai · Tokyo · Kuala Lumpur · Singapore · Sydney

Dr. Anatoliy I. Yashin
Duke Center for Population Health and Aging
Erwin Mill Building
2024 West Main Street
Box 90420
Durham, NC 27705
USA

Dr. S. Michal Jazwinski
Tulane Center for Aging
Department of Medicine
1430 Tulane Ave., SL-12
New Orleans, LA 70112
USA

Library of Congress Cataloging-in-Publication Data

Aging and health (Yashin)
  Aging and health : a systems biology perspective / volume editors,
Anatoliy I. Yashin, S. Michal Jazwinski.
      p. ; cm. -- (Interdisciplinary topics in gerontology, ISSN 0074-1132
; vol. 40)
  Includes bibliographical references and indexes.
  ISBN 978-3-318-02729-7 (hardcover : alk. paper) -- ISBN 978-3-318-02730-3
(e-ISBN)
  I. Yashin, Anatoli I., editor. II. Jazwinski, S. Michal, editor. III.
Title. IV. Series: Interdisciplinary topics in gerontology ; v. 40.
0074-1132
  [DNLM: 1. Aging--physiology. 2. Systems Biology. 3. Aged--physiology.
4. Geriatric Assessment. W1 IN679 v.40 2015 / WT 104]
  QP86
  612.6'7--dc23
                                            2014027396

Bibliographic Indices. This publication is listed in bibliographic services, including Current Contents® and PubMed/MEDLINE.

Disclaimer. The statements, opinions and data contained in this publication are solely those of the individual authors and contributors and not of the publisher and the editor(s). The appearance of advertisements in the book is not a warranty, endorsement, or approval of the products or services advertised or of their effectiveness, quality or safety. The publisher and the editor(s) disclaim responsibility for any injury to persons or property resulting from any ideas, methods, instructions or products referred to in the content or advertisements.

Drug Dosage. The authors and the publisher have exerted every effort to ensure that drug selection and dosage set forth in this text are in accord with current recommendations and practice at the time of publication. However, in view of ongoing research, changes in government regulations, and the constant flow of information relating to drug therapy and drug reactions, the reader is urged to check the package insert for each drug for any change in indications and dosage and for added warnings and precautions. This is particularly important when the recommended agent is a new and/or infrequently employed drug.

© Copyright 2015 by S. Karger AG, P.O. Box, CH–4009 Basel (Switzerland)
www.karger.com
Printed in Germany on acid-free and non-aging paper (ISO 9706) by Kraft Druck, Ettlingen
ISSN 0074–1132
e-ISSN 1662–3800
ISBN 978–3–318–02729–7
e-ISBN 978–3–318–02730–3

Contents

Introduction

Systems biology is a reemerging discipline. Its origins are found in Ludwig von Bertalanffy's general system theory, which eschews reductionism and treats the organism thermodynamically as an open system. A good exposition of this approach is contained in his compendium [1], which is still relevant.

Although general system theory had a significant impact on various disciplines, notably informatics, its consideration in biology, from which it sprung, waned. Systems biology rose again on this substratum around the year 2000. A significant impetus for this was the development of various '-omics', with their capability of generating vast datasets pertaining to cell and organism behavior. The majority of efforts have since been devoted to the generation of networks, and layers of networks, to deduce the multiple interactions of the variables in these datasets.

Another antecedent to the current systems biology is the work of mathematical biologists, whose efforts to model biological processes dynamically feature importantly in some 'strains' of systems biology. Metabolic control analysis comes to mind immediately, as does the literature on the mathematical modeling of the cell cycle. These modeling approaches often incorporate nonlinear functions, and they frequently take into account stochastic elements. These facets are kindred to the consequences of the interaction between components of a system in general system theory. The efforts of both the network systems biologists and the dynamic systems biologists should be juxtaposed to the work of bioinformaticians, who devise methods for manipulating large datasets and cataloging their features.

The systems biology of aging has an even more recent history, although the relevance of the systems approach to aging was already heralded in 1996 [2]. The two sorts of systems biology referred to in the previous paragraph coincide roughly with bottom-up and top-down approaches to the modeling of biological systems. A useful consideration of how these distinct approaches can be profitably integrated has been presented [3]. Most efforts to date attempt to understand the aging process as a determinant of longevity or demise. Little attention has been paid, however, to the emergence of disease and dysfunction as a result of aging, or to the information this emergence has on the biological aging process itself.

The initial idea for this monograph was to explore the frontiers of knowledge connecting aging and health, within a systems biology framework. The crucial importance of this approach lies in the possibility of improving population health by postponing aging or by slowing down individual aging rates. For various reasons, this idea was difficult to realize fully. One reason is that many aspects of the aging process remain unclear and continue to be under intense study, making a discussion of their connections to health perhaps premature. Another reason is that the systems biology of aging is a developing discipline as well, with many new ideas and methods still to appear and to evolve. These factors restricted the scope of this volume and focused it on the foundations and specific aspects of the systems biology of aging, with particular attention to the links between aging changes and diseases of the elderly where corresponding information is available.

The first two chapters introduce the reader to network systems analysis. In the first one by *Tarynn M. Witten*, the author briefly addresses the history of systems biology and introduces the notion of complexity, which manifests itself through nonlinear dynamics, hierarchies and network analysis and can be used to study the intricate and fascinating behaviors of living systems. She suggests treating the biological organism as a network. Then, she explains how network mathematics (graph theory) can provide deeper insight and can even predict potential genes and proteins that are related to the control of organismal life span. The author reviews the history of network analysis at the cellular level and introduces various commonly used network variables. She shows how these variables can be used to predict potential targets for experimental analysis. She also discusses some of the challenges that network methods face.

The second chapter by *Christopher Wimble* and *Tarynn M. Witten* applies the ideas and methods described in the first chapter to concrete examples, using *Saccharomyces cerevisiae* and *Caenorhabditis elegans*. The authors consider possible aging-related changes in a network, which include inactivation of active nodes/activation of inactive nodes (e.g. genes) and loss of connectivity/increase in connectivity. The factors affecting these processes are not considered. The authors show that the network structure determines its vulnerability to possible targeted attacks. Attacks that knock out essential genes disrupt the life span network because the organism dies when an essential gene is knocked out. The authors believe that understanding patterns in network decomposition could lead to early detection of potential neurodegenerative disorders and to potential pharmaceutical intervention at earlier points of disease development.

The third chapter by *Mark Mc Auley* and *Kathleen M. Mooney* focuses on the application of computational systems biology in aging research starting with the rationale for using it for investigating the aging process. The authors discuss alternative theoretical frameworks that can be used to build models of the complex age-related disorders associated with unhealthy aging. The chapter starts with the description of dynamic modeling using differential equations. Then, it incorporates aspects of network analysis and agent-based modeling. Computational modeling is supposed to be an integral component of systems biology, amalgamating with the other techniques discussed in this book to quantitatively represent and simulate biological systems.

The evolutionary theories of aging of biological systems are widely discussed in the literature [4–11]. These theories claim that because aging is largely a postreproductive phenomenon, it should not evolve by natural selection. *Joshua Mitteldorf* believes that aging could be advantageous for stability of ecosystems and hence can be the result of natural selection. The author pays attention to the fact that animals and plants have biological clocks that help to regulate circadian cycles, seasonal rhythms, growth, development and sexual maturity. He puts forth the hypothesis that evolutionarily evolved aging is also clock driven. He focuses on the epigenetic process of DNA methylation, as a clock mechanism, and its relevance to stem cell aging, in particular, in his chapter. Research on the relationship between methylation and aging is still in an early stage, and it has not yet even been proven that alterations of the methylation state are a cause and not simply a product of aging. The hypothesis that the body's age is stored within the cell nucleus as a methylation pattern suggests a program of research and an anti-aging strategy. If validated, this hypothesis would point to a challenging target for medical intervention. Recent results [12] provide additional information for thinking in this direction.

To what extent can insights derived from the systems biology of aging in animal model systems be applied to human aging? *Michael Rose* and his colleagues argue that systems biology of aging might have a different focus in two types of species. The authors provide evolutionary arguments that aging processes taking place in species with rare sexual recombination are quite different from those in which it is frequent. In the species of first type, the systems biology of aging can focus on large-effect mutants, transgenics, and combinations of such genetic manipulations. In frequently recombining species, the systems biology of aging can examine the genome-wide effects of selection.

Many gerontologists have the strong belief that aging is nonprogrammed and provide arguments supporting this view [13]. Many others provide arguments that aging is likely to be programmed [14–16]. Further studies are needed to resolve the issue. *Bruno Cesar Feltes* and his colleagues treat aging as a programmed process and consider it as a continuation of developmental processes. To overcome environmental challenges, the embryo needs to adapt its metabolism in response to environmental fluctuations. Epigenetic programming is responsive to perturbations or imbalances of intrinsic and/or extrinsic factors experienced in utero. Immune system development and aerobic respiration/glucose metabolism processes are modulated during early development. Small changes in developmental mechanisms and adult trait specification that occur during early development might result in significant morphological alterations during later stages. This can promote an adaptive response and influence gene expression patterns, leading to age-associated diseases, such as cancer, osteoporosis and the decline of the immune system. This concept underpins a network approach to aging that provides a framework for the appearance of diseases of aging.

In the chapter that follows, *Arnold Mitnitski* and *Kenneth Rockwood* describe the use of their frailty or deficit index to characterize the state of an aging human. This is a top-down approach that incorporates age-related disease and dysfunction into its derivation. The authors propose that the frailty index can be used as an indicator of an individual's biological age. This index manifested reproducible properties including nonlinear increase with increasing age, higher values in women, strong association with mortality and other adverse outcomes, as well as other properties. Importantly, the authors employ a stochastic dynamics approach to model how the organism recovers as a function of age.

Aging is associated with immunosenescence, and it is accompanied by a chronic inflammatory state which contributes to development of chronic conditions. The chapter by *Verónica Guarner* and *Maria Esther Rubio-Ruiz* shows how low-grade systemic inflammation may be the basis of multiple dysfunctions that evolve during aging, including metabolic syndrome, diabetes, and their cardiovascular consequences. Cardiovascular diseases and endothelial dysfunction are characterized by a chronic alteration of inflammatory function, markers of inflammation, and the innate immune response. Inflammation may thus serve as the integrating factor that makes the frailty index a global measure of system function.

Pharmacologic interventions are believed by many gerontologists as a possibility for slowing down or postponing individual aging processes. A widely discussed target for such interventions is the mTOR (mammalian target of rapamycin) nutrient response pathway. In multicellular organisms, TOR regulates cell growth and metabolism in response to nutrients, growth factors and cellular energy state. Deregulation of TOR signaling alters whole-body metabolism and causes age-related disease. The life-extending effects of dietary restriction in yeast, worms, flies and mice appear to be due largely to inhibition of TOR signaling. There is evidence that TOR may also control aging via modulation of stress-responsive genes and through autophagy. Inhibition of this pathway extends life span in model organisms and confers protection against a growing list of age-related pathologies. In the next chapter, *Simon Johnson* and his colleagues focus their attention on mTOR signaling. The authors inform that some medical interventions affecting this pathway are already clinically approved, and others are under development. Thus, targeting the mTOR pathway is a promising strategy for slowing down the aging rate and improving health of the elderly.

In the following chapter, *Rüdiger Hardeland* discusses melatonin as a systemic integrating agent that interfaces with the environment. A number of studies support the anti-aging properties of melatonin [17, 18]. Melatonin is a derivative of the amino acid tryptophan and widely distributed in food sources, such as milk, almonds, bananas, beets, cucumbers, mustard, and tomatoes. In humans, melatonin is primarily synthesized by the pineal gland, but it is also produced in the gastrointestinal tract and retina. Melatonin and its metabolites are potent antioxidants with anti-inflammatory, hypotensive, cell communication-enhancing, cancer-fighting, brown fat-activating, and blood lipid-lowering effects, and thereby protecting tissues from a variety of insults. Melatonin has been shown to support circadian rhythm, hormone balance, reproductive health, cognition, mood, blood sugar regulation, and bone metabolism, while improving overall antioxidant status and lowering blood pressure. Melatonin may assist in preventing diabetic complications, and improving treatment outcomes in patients with cardiovascular disease and certain types of cancer. Consuming melatonin neutralizes oxidative damage and delays the neurodegenerative process of aging [19]. *Hardeland* here shows that this chronobiotic impinges on multiple physiologic systems with implications for health and disease during aging. The chapter discusses the associations of the loss of melatonin secretion and rhythm amplitudes with aging and development of age-related diseases.

It is well known that a diet rich in plant-based foods has many advantages in relation to the health and well-being of an individual. Much less known is the large contribution of the gut microbiota to this effect. *Denise B. Lynch* and her colleagues expand the discussion of aging, health, and disease to encompass the gut microbiome and its mutual relationship with the host. This relationship goes beyond an uneasy symbiosis implicated in immune-related disorders because the host genome and the microbial ecosystem constitute a supergenome. Thus, this is more than an interaction of the host with the environment with significant consequences for healthy aging.

The penultimate chapter by *Anuradha Chauhan* and colleagues serves as a coda. The authors reprise the history of the systems biology of aging and the different methodological approaches it encompasses. They provide the rationale for using the methods of systems biology in the analyses of the aging of biological systems. They outline the main features of the methodology emphasizing that the structure and functions of the biological systems are investigated by analyzing experimental data through the use of sophisticated mathematical and computational tools, including advanced statistics, data mining, and mathematical modeling. The methodology also includes formulation of working hypotheses, designing new experiments able to prove these hypotheses, and developing computational tools with predictive ability in a biomedical environment. The authors provide several examples that make direct use of the system motifs introduced in previous chapters, and they point to the importance of expanding upon the rudimentary achievements of the systems biology of aging at the present time if we are to intervene in the appearance and progression of age-related disease. The authors believe that the optimal design of biomedical strategies to counteract aging-associated pathologies will require the use of tools and strategies adapted from engineering.

The final chapter by *Vladimir N. Anisimov* addresses the issue of interventions raised again by *Chauhan* and colleagues. He describes experimental studies evaluating effects of biguanides and rapamycin on survival and carcinogenesis in mice paying attention to similarity in the majority of effects of these drugs on patterns of changes observed during normal aging and in the process of carcinogenesis. Anisimov considers whether an antiaging drug is in hand, one that combats age-related disease. The conclusion is that promising leads may already be available.

This book is bound to leave the reader unsatiated. The systems biology of aging is a new field. Although it is based on established methodologies, their application has been relatively limited to date. Furthermore, aging presents problems that are peculiar to it. Some of these peculiarities derive from the forces underlying its evolution. Others are the result of its fundamentally stochastic nature and its heterogeneity among individuals. Its presentation as a set of multiple morbidities and comorbidities only adds to the difficulty. We expect that future research will make use of new concepts and new tools to allow these aspects of aging to be adequately treated. Furthermore, we trust that this volume will stimulate such endeavors.

S. Michal Jazwinski, New Orleans, La.
Anatoliy I. Yashin, Durham, N.C.

References

1 Von Bertalanffy L: General System Theory: Foundations, Development, Applications, rev ed. New York, George Braziller, 1969.

2 Jazwinski SM: Longevity, genes, and aging. Science 1996;273:54–59.

3 Kriete A, et al: Systems approaches to the networks of aging. Ageing Res Rev 2006;5:434–448.

4 de Magalhaes JP, Toussaint O: The evolution of mammalian aging. Exp Gerontol 2002;37:769–775.

5 Bredesen DE: The non-existent aging program: how does it work? Aging Cell 2004;3:255–259.

6 Capri M, et al: Human longevity within an evolutionary perspective: the peculiar paradigm of a post-reproductive genetics. Exp Gerontol 2008;43: 53–60.

7 Goldsmith TC: Aging as an evolved characteristic – Weismann's theory reconsidered. Med Hypotheses 2004;62:304–308.

8 Holliday R: The evolution of human longevity. Perspect Biol Med 1996;40:100–107.

9 Heininger K: Aging is a deprivation syndrome driven by a germ-soma conflict. Ageing Res Rev 2002;1: 481–536.

10 Hughes KA, Reynolds RM: Evolutionary and mechanistic theories of aging. Annu Rev Entomol 2005;50:421–445.

11 Williams PD, Day T: Antagonistic pleiotropy, mortality source interactions, and the evolutionary theory of senescence. Evolution 2003;57:1478–1488.

12 Horvath S: DNA methylation age of human tissues and cell types. Genome Biol 2013;14:R115.

13 Blagosklonny MV: Aging is not programmed: genetic pseudo-program is a shadow of developmental growth. Cell Cycle 2013;12:3736–3742.

14 Jin K: Modern biological theories of aging. Aging Dis 2010;1:72–74.

15 Goldsmith TC: Arguments against non-programmed aging theories. Biochemistry (Mosc) 2013;78:971–978.

16 Goldsmith TC: Aging theories and the zero-sum game. Rejuvenation Res 2014;17:1–2.

17 Sharman EH, et al: Age-related changes in murine CNS mRNA gene expression are modulated by dietary melatonin. J Pineal Res 2004;36:165–170.

18 Acuna-Castroviejo D, et al: Melatonin, mitochondria, and cellular bioenergetics. J Pineal Res 2001; 30:65–74.

19 Pohanka M, et al: Oxidative stress after sulfur mustard intoxication and its reduction by melatonin: efficacy of antioxidant therapy during serious intoxication. Drug Chem Toxicol 2011;34:85–91.

Yashin AI, Jazwinski SM (eds): Aging and Health – A Systems Biology Perspective.
Interdiscipl Top Gerontol. Basel, Karger, 2015, vol 40, pp 1–17 (DOI: 10.1159/000364922)

Introduction to the Theory of Aging Networks

Tarynn M. Witten

Center for the Study of Biological Complexity, Virginia Commonwealth University, Richmond, Va., USA

Abstract

This chapter will briefly address the history of systems biology and complexity theory and its use in understanding the dynamics of aging at the 'omic' level of biological organization. Using the idea of treating a biological organism like a network, we will examine how network mathematics, particularly graph theory, can provide deeper insight and can even predict potential genes and proteins that are related to the control of organismal life span. We will begin with a review of the history of network analysis at the cellular level and follow that by an introduction to the various commonly used network analysis variables. We will then demonstrate how these variables can be used to predict potential targets for experimental analysis. Lastly, we will close with some of the challenges that network methods face. © 2015 S. Karger AG, Basel

In this chapter, we will briefly address the history of systems biology and complexity theory and their use in understanding the dynamics of aging at various levels of biological organization. Using the idea of treating a biological organism like a network, we will examine how network mathematics, focusing on graph-theoretic methods, can provide deeper insight and can even predict potential genes and proteins that are related to the control of organismal life span and perhaps even related to diseases associated with age-related changes within the organism or health span. We will begin with a review of the history of network analysis as related to the study of aging and follow that by an introduction to the various commonly used network analysis constructs. We will then demonstrate how these network variables can be used to further understand and possibly predict potential targets for experimental analysis. Lastly, we will close with some of the challenges that network methods face.

Aging – being old – is defined both biologically and psychosocially [1], 'The geriatric or elderly patient is defined as an individual whose biological age is advanced. By definition, such an individual has one or more diseases, one or more silent lesions in various organ systems.' In addition, physiological changes affect the response to or handling of various medications. Social aspects of aging are also complex, and they include adapting to lessened physical capabilities and often to reduced income and to reduced social network support. For example, many older persons find themselves living alone after decades of marriage, partnership, and/or child rearing. Aging is an intricate spatial and temporal hierarchy of dynamic behaviors that are coupled together in a complex dance across the life span. Thus, aging is a complex, multidimensional, hierarchical process not easily dissected into disjoint subprocesses. How then do we grapple with the problem of understanding such systems?

In the Beginning: Reductionism

Historically, the pursuit of science has taken place by breaking objects apart and subsequently trying to understand how the pieces work at increasingly smaller and smaller levels of organization, the reductionist methodology. It was tacitly assumed that one could just glue the pieces back together and understand the behavior of the unbroken original system. Reductionist methods have been and continue to be widely used to understand biological systems and their dynamics. For example, the early genomic studies of aging identified numerous single genes related to survival [2]. If survival is related to 'aging' and the connections between genes/proteins are known, then perhaps networks of genes/proteins can be constructed that should predict other genes/proteins related to aging. If we understand how these genes and proteins function within an organism, then perhaps we can find ways to extend health span [3], control mortality and morbidity and better treat diseases associated more commonly found in elders of a population. Reductionist science has certainly yielded numerous insights into mechanisms underlying the processes of aging, the control of life span and the dynamics of age-related disease/decline in vitality. We now know many more genes and related proteins that appear to control or to be connected with these processes and we have even identified network pathways of importance [4]. Thus, reductionist approaches have led us part of the way down the path to understanding the processes of life span control. However, as we shall soon see, understanding these systems is not as straightforward as simply gluing genes together to form networks and subsequently gluing networks together to form the whole organism [5].

As we will be making use of a large number of terms, any number of which may be unfamiliar to the readers of this text. We begin by defining terms so that we may all begin with a uniform understanding of the chapter vocabulary and how these concepts apply to the study of biological systems as a whole and 'aging' in particular. We begin by defining the words 'complex' and 'complicated'.

Is a System Complicated or Is It Complex?

The terms 'complicated' and 'complex' are frequently used interchangeably in much the same way that the words 'sex' and 'gender' are now assumed to be linguistically equivalent, though they refer to significantly different conceptual constructs. The same can be said about the words complicated and complex. Given that a system has many parts, a system is said to be *complicated* if infinite knowledge of the behaviors of the system's components allows an experimenter to predict all possible behaviors of the system. For example, a pocket watch would satisfy the complicated but not complex criteria. We can understand the behavior of all of the cogs, wheels and springs in the system and, with some effort, we can arrive at what would be considered reasonable inferences concerning what the watch does and how it works. Breaking apart an organism costs information about how the 'whole' organism functions. This begs the question of whether or not aging can even be reduced to discrete causes, or whether it involves a 'complexity effect' that no single part or collection of parts can fully explain. Systems that lose information in breaking them apart are called 'complex' systems. But what does a complex system actually look like? What might its properties be?

Properties of Complex Systems
If we were to examine a large collection of different complex systems, we would find that complex systems have certain common or unifying characteristics:
- They demonstrate emergent behavior; behavior that cannot be inferred from a linear analysis of the behavior of the components.
- They contain many components that are dynamically interacting (feedback, controllers, detectors, effectors and rules). There is no master controller. The parts interact extensively at their local level with nearest neighbors.
- The components are diverse, thereby leading to a significant diversity of information in the system.
- The components have surrendered some of their uniqueness or identity to serve as elements of the complex system. This is called dissolvence.
- All interactions of the components within the system and the system acting as a component in a higher hierarchy occur locally. There is no action at a distance.
- These interactions take place across a number of scale levels, and they are arranged in a hierarchical structure where fine structure (scale) influences large-scale behavior.
- They are able to self-organize, to adapt and to evolve.

As we can see, complex systems have properties that we do not expect to see in a pocket watch. Complex systems possess additional properties (e.g. control features, feedback loops and branches) that add order, robustness and stability to the system. Complex systems also exhibit an ability to adapt (i.e. evolve) to changing conditions. For example, changes in one free radical-scavenging pathway can up- or downregulate other pathways. Another way to think of complex systems is that they are systems in which the whole is greater than the sum of the parts [6]. Why is this distinction important?

One of the most important properties that distinguish complex systems from complicated systems is the property of emergence. Consider the following examples. Infinite knowledge of a single bird or fish would not allow an experimenter to predict the phenomena of swarming or schooling or the synchronization of firefly lights [7]. Infinite knowledge of a single female's menstrual cycle would not predict cycle locking in a college dorm room. These systems are termed complex [8]. They have 'emergent properties', meaning that a behavior that was not predicted from infinite knowledge of the parts emerges as part of the system's behaviors [9]. Living systems, whether they are cells or ecosystems, do not function like pieces of a jigsaw puzzle. Instead, they are often fuzzy or stochastic, with backup systems and redundancies that belie their true structure. An examination of these systems requires a different conceptual framework. From a Positive Psychology perspective, Maddi [10] makes the argument to '…consider creativity as behavior that is innovative…'. We could easily argue that innovative behavior is emergent behavior, and therefore creativity is an emergent and unpredictable process. Thus, in order to understand complex systems, we must understand them through a reverse engineering perspective rather than a reductionist perspective.

Nonlinear Dynamics and Aging
By the early 1800s, studies of biological systems, ecosystems in particular, were observed to demonstrate a variety of nonlinear behaviors; particularly oscillations, apparently chaotic time series and radical behavioral changes that could not be explained by traditional reductionist constructs [7]. From the early work of von Bertalanffy [11] and many others emerged the concepts of systems dynamics and systems theory as applied to a variety of living systems. Very early on, ecologists saw the value of systems theoretic approaches in understanding the complex ecological systems with which they worked. However, it was not until the work of Rosen [12] on MR systems and the subsequent work of May [13] and others who began to write about simple nonlinear models with complex dynamics (these are classic papers) that we began to see the emergence of previously described nonlinear phenomena such as chaos.

Nonlinear systems theory and multifractal analysis have already been used to understand fall safety in elders, frailty in the elderly, wandering in community-dwelling older adults, understanding interactions of geriatric syndromes and disease and in understanding the brain structures of Alzheimer patients. Network analytic methods have been used to construct longevity gene-protein networks and to predict potential gene targets of importance to longevity and perhaps to pharmacological intervention. Consequently, systems biology is now emerging as a powerful paradigm for understanding networks of longevity genes and proteins. With the sequencing of the human genome, massive amounts of data have been generated by the 'omics' disciplines over the past twenty years; including genomics, proteomics, metabolomics, transcriptomics, and interactomics. An excellent discussion of complex systems dynamics and nonlinear dynamics may be found in Strogatz [7].

The application of the pantheon of mathematical and computational tools of systems biology has the potential to help transform the massive amounts of data into useful information that can be used to understand the biomedical processes associated with human disease and potentially how they relate to the dynamics of aging. By integrating omic data with the identification of critical networks and pathways associated with specific diseases of age and with vitality and longevity, greater understanding of these biological processes can be achieved. This enhanced understanding can help biomedical researchers design new and better approaches to treat or to manage the diseases of age and to help develop strategies to promote enhanced vitality and longevity, what is more currently known as health span. As the 'baby boomers' move into their 60s and 70s, increased demand for care for the diseases of age and for approaches to enhance vitality and promote longevity means that new and improved remedies and interventions will be required. Consequently, a systems approach to the study of aging and its processes offers promise as a means of attaining potentially significant gains in the management and treatment of age-related diseases.

On the one end of the spectrum, we have reductionist methods that have allowed us to see into the organism and determine genes associated with life span. At the other end of the spectrum, we have holistic or complexity theoretic methods that allow us to probe an organism with minimal perturbation. Where does Systems Biology fit in?

The Emergence of Systems Biology

Systems science takes a middle ground approach, neither reductionist nor holistic [14]. It attempts to look at the parts and it tries to glue them back together under the assumption that whatever complexity-related information is lost does not profoundly affect understanding the behavior of the system. Like a jigsaw puzzle, pieces are linked into chains that are then used to form small networks from which a picture of the process begins to emerge. While it was often possible to gain insights into the system behavior by gluing parts back together, for many systems it just did not work. This was particularly true for living systems in all of their forms and beauty. Life, it seems, was far more 'complex' than had been thought [15]. However, given the early lack of data on the pieces of biological systems and the minimal knowledge on how they were connected, it seemed that the only obvious approach was to try to glue pieces into potential networks, then glue the networks into hierarchies and finally see what results were obtained. The initial developments, particularly as applied to studies in gerontology and geriatrics, evolved from the idea of building reliable biological organisms.

An Example: Reliability Theory

One of the earliest aging-related uses of systems biological approaches was the use of what is now called *reliability theory*. The constructs of reliability theory emerged from the 1950s *gedankt* experiments of the computer scientist John Von Neumann [16]. Von Neumann's interest [see 17] was in how one would go about building a reliable biological organism out of unreliable parts. This question led to the development of the field of reliability theory and the subsequent adaptation of the field of reliability theory to become what is now known as the field of survival theory. Until the thought experiments of von Neumann, the concept of reliability had not been well defined.

Von Neumann's argument proceeded as follows. He began by defining the concept of the *conditional instantaneous failure rate*, denoted by $\lambda(t)$. We interpret this as follows. The condition is that the failure has not occurred at time t given that the organism has survived until time t. With this in mind, we may then define the reliability $R(t)$ of an organism as the probability of no failure of the organism before time t. If we let $f(t)$ be the time to (first) failure (this is the same as the failure density function), then the reliability $R(t)$ is given by $R(t) = 1 - F(t)$, where

$$F(t) = \int_0^t f(\tau)d\tau \text{ [18]}.$$

How do we actually obtain an equation for the reliability $R(t)$? We do this as follows. Suppose we ask what is the reliability $R(t + \Delta t)$ where Δt is a small time increment. In other words, suppose that we know the reliability of the organism at time t and we want to know the organism's reliability at a small time increment Δt later than time t. In order for the organism to be operational at time $t + \Delta t$, the organism must have been operational until at least time t and then not have failed in the time interval $(t, t + \Delta t)$. We can express this mathematically as follows. The reliability $R(t + \Delta t)$ is given by

$$R(t + \Delta t) = R(t) - \lambda(t)R(t)\Delta t \tag{1}$$

Reading equation 1, we see that to be functional (operational) at time $t + \Delta t$, the organisms had to be functional at time t [denoted by the reliability term $R(t)$ on the right hand side of the equation]. Next, we have to subtract out all of the items that failed in the time interval $(t, t + \Delta t)$; given by the second term on the right hand side of equation 1. What remains after this subtraction is all of the organisms or items that remain functional at time $t + \Delta t$. A bit of algebraic rearrangement and we have

$$\frac{R(t + \Delta t) - R(t)}{\Delta t} = \lambda(t)R(t) \tag{2}$$

It follows that letting $\Delta t \to 0$ (remembering our calculus), equation 1 becomes the simple differential equation given by

$$\frac{dR(t)}{dt} = -\lambda(t)R(t) \tag{3}$$

Thus, if we can specify the form of the function $\lambda(t)$, we can solve for $R(t)$. The literature in these fields often uses the term 'failure rate function' interchangeably with the term 'hazard' function. For those readers who have dabbled in demography or survival analysis, these constructs should seem quite familiar. Most people are familiar with either the Gompertz mortality rate (hazard rate/failure rate) $\lambda(a) = h_0 e^{\gamma a}$ where h_0 and γ are parameters typically estimated from population data for a given organism. Given the large literature on different mortality rate functions and their applicability to population modeling, we direct the interested reader to the relevant literature in that field. An excellent starting place may be found in Carnes et al. [20].

Systems biology, as applied to the biology of aging was simultaneously and independently originated by Doubal [21], Gavrilov and Gavrilova [22], Koltover [23], Witten and Bonchev [24], in the mid-to-late 1980s. Additional application of network theory to aging may be found in Kirkwood [25] and more recently in Qin [26] and Wieser et al. [53]. These papers focused on two application areas, genetic and general network theoretic applications. The thinking was that biological systems, particularly cellular systems, could be thought of in the same way as networks with interconnected parts that had certain failure rates. The death of the organism, and hence its life span, could be thought of as a network failure. The discipline of reliability theory, coupled with network analysis/graph theory, allowed these researchers to hypothesize certain network structures and to subsequently calculate failure curves for those network structures. In a number of cases, the shapes of the network survival curves mimicked the population survival curves seen in real biological populations, suggesting that reliability theoretic approaches, coupled with network assumptions, might have something to offer in understanding aging at a demographic level. This is because concepts of reliability have direct analogs to the longevity and lifespan of an organism. The most obvious one is that life span can be thought of as 'the time to failure' of an organism. If death can be viewed as a failure, then there is a natural linkage between survival and reliability. Thus, the ideas of reliability mutated and the mutation became what we now know as the field of survival theory. Reliability theory allows researchers to predict the age-related failure kinetics for a system of given architecture (reliability structure, network, graph) and given reliabilities of its components.

During the past decade, with the increase in pathway 'omic' information, there has been an increased use of complexity theoretic and systems biological tools and techniques to address putting the pieces of cellular networks back together so that their network properties can be better understood [27]. These methods have also been applied to understanding the dynamics of cellular and molecular aging networks [28–30]. The systems biology approach has begun to allow researchers to understand the effects of multiple complex interactions in these aging networks, thereby further advancing our understanding of how longevity, vitality, and aging-related diseases may be managed. While reductionist approaches are still important, systems biology methods and complex systems theory constructs such as dynamical systems theory, network analysis, fractal dynamics, multi-level computational modeling and swarm theory can extract real information out of terabytes of data, and the role of systems biology and complex systems theory is now emerging as the front-running paradigm for understanding molecular and cellular networks of longevity genes and proteins. How then do we begin to understand networks?

Networks and Graphs

Much of the early work in graph theoretic applications to aging was based upon assumptions about how the genes were connected, as large databases of genes and networks simply did not exist. As available biological data increased, theoretical approaches, though more rigorously tied to experimental data, still struggled with questions around accuracy and reliability of the 'omic' data being used. As the data became cleaner, it became possible to connect single life span-related genes and proteins into component networks. Other networks that controlled heat shock and other biological processes began to be identified. And now, with GWAS methods, we can begin to tie multiple cellular networks to the longevity gene networks. These networks could then be represented as mathematical structures called graphs [31, 32]. These graphs could then be analyzed using the techniques of mathematical graph theory, particularly in light of the recent developments in network topology [33] and its implication for small-world theory [34], scale-free theory [35], redundancy [36], robustness [37], frailty [38], evolvability [39] and resilience [40] of the original biological network.

From Data to Graphs

A graph G is simply a set of *nodes* or *vertices* $n_1, n_2, ... n_G$ and *edges* E_{ij} that connect some or all of the nodes to each other. From a biological perspective, we can consider the nodes to be genes or proteins and the edges as paths between them. We represent the overall network connectivities in a matrix format called the *adjacency matrix* which we denote with the symbol A. The elements of A are denoted a_{ij} and are simple; if node n_i is connected to node n_j, we enter the number one in the $(i,j)^{th}$ element of the matrix A, otherwise we enter a zero. Observe that if n_i is connected to n_j, then it follows that n_j is connected to n_i so that the matrix A is a symmetric matrix. In the case where there are multiple edges connecting the same nodes, we enter the number of edges. Thus, if two different edges connect n_i to node n_j, we would enter the number two. Nodes that are not connected to anything in the graph G are called *islands*. The edges can have *weights*, denoted w_{ij}, assigned to them where, for example, the weight value may correspond to a rate of reaction. An edge E_{ij} can also have a direction assigned to it. For example, if E_{12} represents the edge between nodes n_1 and n_2, we might denote the fact that n_1 is upstream of n_2 by $E_{1 \rightarrow 2}$. An edge that does not have any direction assigned to it is said to be *undirected*, whereas edges that have a direction assigned to them are called *directed* edges. Note that we can have other types of edges in a network. For example, *multi-edges* are multiple edges between nodes, and *self-edges* occur when a node is connected to itself. With this simple set of definitions, we have some powerful tools with which to investigate the structure of a network and how it might inform us about the biological dynamics of the overall network. We begin with the concept of connectivity.

Network Structures and Connectivity

It is natural to conclude that the more edges going in and out of a node, the more likely that the given node is going to be of importance to the network. *Hubs* or nodes with large numbers of connections are known to play central roles in keeping complex networks connected. This is important when we consider, in an upcoming section, the concepts of *robustness*, *resilience* and *frailty* of a network. The number of connections k_i going in and out of a node n_i is called the *connectivity* or *degree* k_i of the ith node. Sometimes you will see the degree of a node expressed using $d(n_i)$. In mathematical terms

$$k_i = d(n_i) = \sum_{j=1}^{N} a_{ij},$$

where N is the number of nodes in the network and a_{ij} is the $(i,j)^{th}$ element of the adjacency matrix A. Computing the connectivity of a large set of nodes leaves us with nothing more than a frequency table, and it is hard to interpret this string of numbers k_i, particularly if the number of nodes in the network is large. In order to assist us in understanding the connectivity structure of the network, we create a connectivity plot. To do this, we first count the number of nodes with a given connectivity k, where the connectivity varies from zero to the maximum connectivity value. The number of nodes with a given connectivity k is called the frequency of that connectivity and is denoted $f(k)$. Next, plot the frequency $f(k)$ versus the connectivity k.

Studies of the statistical behavior of various network structures [41, 42] have shown that networks can have a small variety of overall topologies [43]: random, regular, small world and scale free. Moreover, each of these network topologies has a classic pattern form for its degree distribution plot. *Random networks* are just what you would imagine them to be; nodes are randomly connected to each other. *Regular networks* can be thought of as lattices where there is a repetitive pattern of connections such as a grid. Small-world and scale-free networks are of greater interest because they have some fascinating underlying properties. Moreover, many real-world networks can be shown to be small world or scale free [34]. A *small-world network* can be described as a network in which most nodes are not neighbors of one another, but most nodes can be reached from every other node by a small number of hops or steps [34]. A *scale-free network* may appear to be a random network; however, in a scale-free network the links between the nodes are preferentially attached to the most highly connected nodes,

Fig. 1. Illustration of a sample connectivity or *degree distribution* plot for the network. See Witten and Bonchev [24] for more details. The rhombs represent the complete distribution. The squares are the data points binned into groups of three. The black solid line is the nonlinear regression line. Results are significant at p < 0.05.

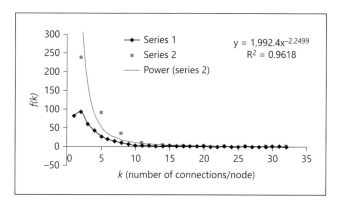

thereby creating a greater frequency of links connected to a smaller number of nodes [35]. Because scale-free networks are ubiquitous and highly relevant to our discussion, let us look at them a bit more closely. In examining figure 1, we see that not all nodes in the network have the same number of edges. If we divide the y-axis in figure 1 by the total number of nodes in the network, call that N, then $f(k)/N$ represents the probability $P(k)$ that a randomly selected node has exactly k edges. In a randomly connected graph, the edges are placed at random, and one can show that the majority of the nodes will have approximately the same connectivity which is close to the *average connectivity <k>*. In fact, it has been shown that the connectivities k in a random network follow a Poisson distribution with a peak at *<k>*.

What became interesting is that, for larger networks like gene, protein and metabolic networks, these networks did not follow the traditional Poisson probability distribution. Rather, they followed a probability distribution where the connectivity probability $P(k)$ was a power law of the form

$$P(k) = Bk^{-\gamma} \tag{4}$$

Observe that since $P(k)$ is a probability, when we sum over all of the values of k, the result had better add up to one. Thus, the parameter value of B is chosen so that this is true. We will not get into all of the varied aspects of scale-free networks [31, 33]. However, how can we determine if we have a scale-free distribution?

Power Plots and Scale-Free Networks
We observe that if we take the log of both sides of equation 4 the better the fit, the more linear the plot should be. Thus, networks whose connectivity structure follows a power law of the form $f(k) = Bk^{-\gamma}$, where B and γ are parameters to be estimated and should look like negative slope lines if they are scale free. The simplest way to estimate the parameters is to perform a linear regression on the log-log transformed $f(k)$ versus k data, dropping the $k = 0$ data point because there are no connectivities. The more linear this curve, the more the connectivity behaves like a power law.

Why should we even care about degree distributions and power law networks in the first place? As we noted earlier, scale-free networks are ubiquitous in living and even nonliving systems. Scale-free networks are also special in that they are built in a unique way. To build a scale-free network, you start off with a set of N nodes in which each node in the network is connected to all of the other nodes. Next, to add a new node, you make k connections to existing nodes in the network. However, whether a new node m is connected to an already existent network node n_i is determined by the degree of the given node n_i; the greater the degree of n_i the more likely m is going to be connected to n_i. In other words, the probability that node m will be connected to node n_i is given by

$$P(n_i) = \frac{d(n_i)}{\sum_j d(n_j)} \tag{5}$$

Notice that this connectivity algorithm means that if you are already very tightly connected in the network, then you are more likely to get even more connected in the network. Many scale-free networks have an exponent $\gamma \approx 3$. However, the exponent value very much depends upon the rule used for the probability of new node connection. Equation 5 is a very simple example. Given the large number of biological networks, particularly at the cellular level, that have been shown to be small-world formations, this suggests that tendency to create small-small world networks is a natural evolutionary pathway.

Categorizing Small-World Networks
Because of the unique nature of scale-free networks, a log-log connectivity plot is enough to let you know if you are dealing with a scale-free network or not. However, this does not work for other network forms. Because many biological systems also demonstrate small-world network behavior [34], we briefly examine how to determine whether or not a network is a small-world network. Remember, a working definition of a small-world network is a network in which most of the neighbors of a node are neighbors themselves (think regular network here, lattice structure for example). However, in addition to this property, the average number of connections between two chosen random nodes in the network n_i and n_j is small (similar to the properties of a randomly connected network). To help characterize small-world networks, we introduce a few new network descriptors. The first is the *average path length* of a network. Path length is the distance or number of edges between two nodes in the network. So, choose two random nodes in the network, figure out all of the different paths between them and count the number of edges in each of the paths. Then compute the average number of edges and you have the average path length. We can use the idea of path length to construct the *minimum path length* between node n_i and node n_j and denoted ℓ_{ij} and the average minimum length of a network as $<\ell>$ using the same ideas as the average path length.

Node-Node Connectivities

Another common network term is the *centrality* of a node. Centrality is a measure of the 'position' or relative importance of a node in a network. In the literature, there are four main measures of centrality of a node: *degree centrality, betweenness centrality, closeness centrality* and *eigenvector centrality*. From an aging-related perspective, understanding node centrality of the nodes in a network could lead to potential targets for pharmaceuticals that might help hinder disease progression or extend life span. The simplest of the centrality measures is degree centrality. *Degree centrality* of a node is defined by $C_D(n_i) = d(n_i)$. In other words, the degree centrality of a node n_i is simply the number of edges that are connected to the given node n_i. Obviously, this measures the chance that a given node n_i in the network will receive something flowing along the network. In the case where the graph is directed, or we know the flow along the edges (upstream, downstream), we can define two new concepts C_{inD} and C_{outD} as the number of edges going in and out of n_i. These are called *indegree* and *outdegree*, respectively. *Closeness centrality*, denoted $C_C(n_i)$ is the idea that the more central that a node is in a network, the lower its total distance is to all of the other nodes in the network. In other words, if a node n_i is very close to all of the nodes, it should take a small number of edges to get to every other node in the network. From a biological perspective, closeness can be thought of as a measure of how long it will take to send a chemical or other biological signal out from n_i to all of the other nodes in the network. *Betweenness centrality*, denoted $C_B(n_i)$ looks at how often, in a network, a given node n_i acts as a bridge along the shortest path between two other nodes. From a biological perspective, knocking out a node with high betweenness centrality would force a signal to reroute itself along a path that was not the shortest path. Lastly, *eigenvector centrality*, denoted $C_E(n_i)$ is a measure of the 'influence' of a node in a network. Here, the idea is that not all connections between nodes are equal. That is, if a node is influential and it is connected to another node, it is likely that it will have more influence on that node than a node that is not that influential. An excellent discussion of the various concepts of centrality can be found in Opsahl et al. [44].

Clusters and Hierarchies

Earlier on, we mentioned the concepts of clustering. Many biological networks, metabolic networks and protein interaction networks demonstrate both clustering and scale-free properties [45]. When examining network structures of this class of networks, we find that they are often modular and *hierarchical* in nature. That is, networks that exhibit the combination of small-worldness and clustering appear to be built out of modules that are themselves networks. One measure of the intrinsic hierarchical nature of a network is to make use of the mathematical result that deterministic scale-free networks that are hierarchical tend to have a clustering coefficient that goes as $C(k) \approx k^{-1}$. That is, if a node n_i has k connections, then its clustering coefficient is approximately k^{-1}. Thus, the higher a node's degree, the smaller its clustering coefficient. Moreover, the larger

k gets, the more likely the clustering coefficient of the given node behaves as k^{-1} studies of many biological systems have, indeed, shown that the networks demonstrate modularity [27].

Robustness, Resilience and Frailty

We all have an intuitive idea of what robustness, resilience and frailty mean. From an intuitive perspective, resilience can be defined as the ability of a system, when perturbed, to return to its original state of operation [40]. Some people loosen the definition to allow the system to return to a state of operation that is close to the original state of operation, where closeness is defined in such a way that the system is still functional as if it were still in its original state. Like most of the terms that we have been using, resilience is a complexity-related concept. For example, a system can take a short time or a long time to return to its operational zone. Are both of these the same degree of resilience? Surely not! A system can be perturbed for a fixed length of time and then the perturbation stops. What if the return to normalcy time depends upon the length of the perturbation? Are systems that return faster more resilient than ones that take longer to return? Can resilience be used up or built up? Thus, the term resilience encompasses a number of facets, most of which are ignored or tacitly assumed when talking about the subject of resilience. What we need to understand is that resilience is a system response property that allows the system to compensate after it has been perturbed. Since it is a global system property, complexity theory teaches us that it can have unpredictable outcomes due to its inner complexity. Bonanno et al. [46] point out that there are many 'independent predictors of resilient outcomes'. This suggests two things. First, it suggests that resiliency analysis requires nonlinear methods in order to more effectively represent it. Second, it suggests that the human 'resilience system' may be built with some form of redundancy/back-up system, some form of alternative and/or compensatory pathways in case some portion of the resilience system fails. Notice that the constructs of backup and redundancy tie back to our discussion on reliability of network systems [47]. This type of organizational structure suggests that the human 'resilience system' may have a fractal dimension that lies in what is often called the 'robust to attack' domain. That is, the resilience system may have evolved in such a way that it is not frail; not easily vulnerable to attack and/or perturbation. If the system is fractal in nature, then this also suggests the various paths to prototypical outcome trajectories [46].

In the previous discussion, we noted that resilience is a measure of the system's ability to return to an operational space upon perturbation. The fact that the system was able to be perturbed indicates that it was not able to resist the forces of perturbation. This brings us to the concept of robustness. There are many definitions for *robustness*, and they are all context dependent. In one sense, robustness and resilience are opposite concepts. Robustness can be viewed as resistance to perturbation, and resilience is the ability to return to one's original state, or – in the case of positive psychology – to subsequently '[achieve a] positive adaptation despite major assaults on the developmental process' after a perturbation. Thus, robustness to stress implies that it is hard for stress to move a person off their current trajectory (life course), while resilience says that if the individual is moved away from the life course trajectory then how long, if you will, does it take to get back to the original trajectory or to a trajectory that will serve as an acceptable surrogate for the original trajectory.

Systems that are robustly designed are often more difficult to study because they have built in ability to resist perturbation and, if they blow a circuit or an operational unit, they often have backup systems that can keep the living system (or nonliving system) operating in a functional way [48]. Robustness, or 'resistance to perturbation' can be enhanced by developing the client's strengths. Thus, the construct of robustness may be seen as an emergent consequence of a system design that has built in buffers against assaultative pressures. Thus, like resilience, robustness is a global catch-all term designed to describe a system's ability to defend itself against perturbation. Because robustness represents a system's ability to resist perturbation, robustness could also represent a threat to evolvability and adaptability.

Evolvability [52] is the system's ability to alter itself in response to changes in external forces in such a way as to allow the system to continue to function – adaptive evolution. A living system is said to be evolvable if it can acquire novel functions through genetic change, functions that help the organism survive and reproduce. Because the concept of evolvability originally arose in the field of organismal evolution, it carries with it the ideas associated with genetic evolution in the face of exogenous pressures. However, it is not unreasonable to think of living systems such as organizations as having constructs that are equivalent to genes and to ask how an organization might evolve in the face of external pressures such as economic stress. Thus the important nodes in the hierarchy can change as can the connections [39]. These changes can lead to new emergent dynamics that could be considered adaptations to the exogenous stressors or forces.

Conclusion

Closing Thoughts

In this chapter, we reviewed the literature on systems biology in aging from a historic as well as current perspective. We discussed how systems biology of aging emerged from the dynamic interplay of reductionist methodological approaches and the need to address concepts of complexity theory that began to develop in the early 1970s. We showed how these considerations and the perspective of considering a biological system as a network allowed scientists to use graph-theoretic methods to begin to understand the impact of an organism's network structure on its behavior. We then examined some basic concepts of network theory and how they apply to studying cellular aging systems. We discovered that many longevity-related cellular networks have small-world and/or scale-free network properties. Further, it appears that these networks are also very modular in nature. It has been pointed out that this suggests that modularity is important, not because modules exist as somewhat independent subnetworks, but rather because they are combined so that they are tightly connected to each other. These modules are then used to build higher-level network components and networks themselves. We raised questions around what happens when networks 'age'. We saw that this raises questions around network robustness, frailty and susceptibility to various forms of attack, and we examined how network structure may or may not make a network more vulnerable to perturbation, thereby potentially reducing its resilience/robustness and potentially making it more frail. In this chapter, we have focused solely upon graph-theoretic approaches in systems biology. There are many other systems methodological approaches to understanding systems biology of aging. For example, Kriete et al. [49] use a rule-based systems approach to modeling aging systems. Albert et al. [19] use a Boolean network simulation approach. And Huang et al. [50] use a machine learning approach.

Future Directions

The bulk of network aging studies are based upon a snapshot of a single time point in the organism's life span, a cross-sectional picture of the network. In order to really understand how an organism ages (how its networks age), we must better understand how networks evolve over the life span of the organism. There is no reason to believe that network nodes remain active throughout the whole life span, nor should we assume that network edges remain present throughout the life span of an organism. Longitudinal network analyses are now needed. Determining the weights and directions of network connections in aging-related network structures is now important so that accurate simulations of the network dynamics can be developed [51].

In the following chapter, we apply some of these concepts to actual aging networks and illustrate for the reader how to make some of the calculations we discussed in this chapter. Due to chapter restrictions, many relevant references were left out of both this chapter and the following chapter. For this, the author can only apologize for the citation choices made. To compensate for this, I am providing an exhaustive reference list for both chapters. This reference bibliography may be found at http://www.people.vcu.edu/~tmwitten.

Acknowledgements

The author would like to thank many individuals for their respective support and collaborative kindnesses during her career path. In alphabetical order I would like to acknowledge my colleagues and friends: Danail Bonchev, Bruce Carnes, Caleb Finch, Ari Goldberger, Leonard Hayflick, S. Michal Jazwinski, Tom Johnson, Vitaly Koltover, Andres Kriete, George Martin, Ed Masoro, Robert May, Robert Rosen, Bernie Strehler, F. Eugene Yates and B.P. Yu.

References

1 Aronheim JC: Handbook of Prescribing Medications for Geriatric Patients. New York, Little, Brown & Co, 1992, p ix.

2 Miller RA: Genes against aging. J Gerontol A Biol Sci Med Sci 2012;67A:495–502.

3 West GB, Bergman A: Toward a systems biology framework for understanding aging and healthspan. J Gerontol A Biol Sci Med Sci 2009;64:2005–2008.

4 GenAge. http://genomics.senescence.info/genes/.

5 Wong SL, Zhang LV, Tong AH, Li Z, Goldberg DS, King OD, Lesage G, Vidal M, Andrews B, Bussey H, et al: Combining biological networks to predict genetic interactions. Proc Natl Acad Sci USA 2004;101:15682–15687.

6 Gorban A, Petrovskii S: Collective dynamics: when one plus one does not make two. Math Med Biol 2011;28:85–88.

7 Strogatz SH: Sync: How Order Emerges from Chaos in the Universe, Nature and Daily Life. New York, Hyperion Books, 2003.

8 Sumpter DJT: Collective Animal Behavior. Princeton, Princeton University Press, 2010.

9 Sulis W: Archetypal dynamics, emergent situations and the reality game. Nonlinear Dynamics Psychol Life Sci 2010;14:209–238.

10 Maddi SR: Building an integrated positive psychology. J Pos Psychol 2006;1:226–229.

11 Von Bertalanffy L: The theory of open systems in physics and biology. Science 1950;111:23–29.

12 Rosen R: A relational theory of biological systems. Bull Math Biophys 1958;20:245–260.

13 May RM: Simple mathematical models with very complicated dynamics. Nature 1976;261:459–467.

14 Marijuán PC: Bioinformation: untangling the networks of life. Biosystems 2002;64:111–118.

15 Janecka IP: Cancer control through principles of systems science, complexity and chaos theory: a model. Int J Med Sci 2007;4:164–173.

16 Von Neumann J: Probabilistic logics and the synthesis of reliable organisms from unreliable parts; in Shannon CE, McCarthy J, Ashby WR (eds): Automata Studies. Princeton, Princeton University Press, 1956.

17 von Neumann J: Probabilistic logics and the synthesis of reliable organisms from unreliable components; in Shannon C (ed): Automata Studies. Princeton, Princeton University Press, 1956.

18 Kalbfleisch JD, Prentice R: The Statistical Analysis of Failure Time Data. New York, John Wiley & Sons, 2002.

19 Albert I, Thakar J, Li S, Zhang R, Albert R: Boolean network simulations for life scientists. Source Code Biol Med 2008;3:16.

20 Carnes B, Staats D, Vaughn M, Witten TM: An organismal view of cellular aging. Med Longev 2010;2:141–150.

21 Doubal S: Theory of reliability, biological systems and aging. Mech Aging Dev 1982;18:339–353.

22 Gavrilov LA, Gavrilova NS: Reliability theory of aging and longevity; in Masoro EJ, Austad SN (eds): Handbook of the Biology of Aging, ed 6. San Diego, Academic Press, 2006, pp 3–42.

23 Koltover VK: Reliability concept as a trend in biophysics of aging. J Theor Biol 1997;184:157–163.

24 Witten TM, Bonchev DG: Predicting aging/longevity-related genes in the nematode *C. elegans*. Chem Biodivers 2007;4:2639–2655.

25 Kirkwood TB: Network theory of aging. Exp Gerontol 1997;32:395–399.

26 Qin H: A network model of cellular aging. arXiv1305.5784v1 (q-bio.MN).

27 Ravasz E: Detecting hierarchical modularity in biological networks; in McDermott J, Samudrala R, Bumgarner RE, Montgomery K, Ireton R (eds): Computational Systems Biology: Methods in Molecular Biology. New York, Humana Press, 2009, pp 145–160.

28 Kirkwood TB: Systems biology of ageing and longevity. Philos Trans R Soc Lond B Biol Sci 2011;366:64–70.

29 Kriete A, Lechner M, Clearfield D, Bohman D: Computational systems biology of aging. Wiley Interdiscipl Syst Biol Med Rev 2011;3:414–428.

30 Leonov A, Titorenko VI: A network of interorganellar communications underlies cellular aging. IUBMB Life 2013;65:665–674.

31 Dehmer M, Emmert-Streib F (eds): Analysis of Complex Networks: From Biology to Linguistics. Weinheim, Wiley-VCH, 2009.

32 Newman MEJ: The structure and function of complex networks. SIAM Rev 2003;45:167–256.

33 Barabási A-L, Oltvai ZN: Network biology: understanding the cell's functional organization. Nat Rev Genet 2004;5:101–115.

34 Gallos LK, Makse HA, Sigman M: A small world of weak ties provides optimal global integration of self-similar modules in functional brain networks. Proc Natl Acad Sci 2012;109:2825–2830.

35 Reed WJ: A Brief Introduction to Scale-Free Networks. Natural Resource Modeling. London, Chapman & Hall/CRC Press, 2006, pp 3–13.

36 Albert R, DasGupta B, Gitter A, Gursoy G, Paul P, Sontag E: Computationally efficient measure of topological redundancy of biological and social networks. Phys Rev E 2011;84:036117.

37 Kriete A: Robustness and aging – a systems-level perspective. Biosystems 2013;112:37–48.

38 Stromberg SP, Carlson J: Robustness and fragility in immuno-senescence. PLoS Comput Biol 2006; 2:e160.

39 Whitacre JM, Bender A: Degeneracy: a design principle for achieving robustness and evolvability. J Theor Biol 2010;263:143–153.

40 Jackson S: Architecting Resilient Systems. Hoboken, John Wiley & Sons, 2010.

41 Barabási A-L, Albert R: Statistical mechanics of complex networks. Rev Mod Phys 2002;74:47–97.

42 Képès F: Biological Networks. New Jersey, World Scientific Publishers, 2007.

43 Jeong H, Tombor B, Albert R, Oltvai ZN, Barabási A-L: The large-scale organization of metabolic networks. Nature 2000;407:651–654.

44 Opsahl T, Agneessens F, Skvoretz J: Node centrality in weighted networks: generalizing degree and shortest paths. Soc Networks 2010;32:245–251.

45 Jeong H, Mason SP, Barabasi AL, Oltvai ZN: Lethality and centrality in protein networks. Nature 2001; 411:41–42.

46 Bonanno GA, Westphal M, Mancini AD: Resilience to loss and potential trauma. Annu Rev Clin Psychol 2011;7:511–535.

47 Guimerá R, Sales-Pardo M: Missing and spurious interactions and the reconstruction of complex networks. Proc Natl Acad Sci 2009;106:22073–22078.

48 Pradhan N, Dasgupta S, Sinha S: Modular organization enhances the robustness of attractor network dynamics. 2011. http://arxiv.org/abs/1101.5853v1.

49 Kriete A, Bosi WJ, Booker G: Rule-based cell systems model of aging using feedback loop motifs mediated by stress reponses. PLoS Comput Biol 2010; 6:e1000820.

50 Huang T, Zhang J, Xu Z-P, Hu L-L, Chen L, Shao J-L, Zhang L, Kong X-Y, Cai Y-D, Chou K-C: Deciphering the effects of gene deletion on yeast longevity using network and machine learning approaches. Biochimie 2012;94:1017–1026.

51 Albert R, Wang R-S: Discrete dynamic modeling of cellular signaling networks; in Johnson L, Brand L (eds): Methods in Enzymology 476: Computer Methods B. New York, Academic Press, 2009.

52 Kirschner M, Gerhart J: Evolvability. Proc Natl Acad Sci USA 1998;95:8420–8427.

53 Wieser D, Papathenodorou I, Ziehm M, Thornton JM: Computational biology for ageing. Philos Trans R Soc Lond B Biol Sci 2011;366:51–63.

Tarynn M. Witten
Center for the Study of Biological Complexity
Virginia Commonwealth University, PO Box 842030, 1000 West Cary Street
Richmond, VA 23284 (USA)
E-Mail tmwitten@vcu.edu

Yashin AI, Jazwinski SM (eds): Aging and Health – A Systems Biology Perspective.
Interdiscipl Top Gerontol. Basel, Karger, 2015, vol 40, pp 18–34 (DOI: 10.1159/000364925)

Applications to Aging Networks

Christopher Wimble · Tarynn M. Witten

Center for the Study of Biological Complexity, Virginia Commonwealth University, Richmond, Va., USA

Abstract

This chapter will introduce a few additional network concepts, and then it will focus on the application of the material in the previous chapter to the study of systems biology of aging. In particular, we will examine how the material can be used to study aging networks in two sample species: *Caenorhabditis elegans* and *Saccharomyces cerevisiae*.　　　　© 2015 S. Karger AG, Basel

In the previous chapter, we addressed the importance of understanding how complexity theory, as manifested through nonlinear dynamics, hierarchies and network analysis, can be used to study the intricate and fascinating behaviors of living systems. We discussed why reductionism, while it has its uses, also causes us to lose important information about the behavior of a system because breaking a system apart costs us information about how the 'whole' organism functions. We discussed the difference between a complicated system and a complex system, and we then detailed the core properties of a complex system. We pointed out that if we were to examine a large collection of different complex systems, we would find that complex systems have certain common or unifying characteristics. We argued that complex systems have 'emergent properties' meaning that a behavior that was not predicted from infinite knowledge of the parts emerges as part of the system's behaviors. Living systems, whether they are cells or ecosystems, do not function like pieces of a jigsaw puzzle. Instead, they are often fuzzy or stochastic, with backup systems and redundancies that belie their true structure. And we pointed out that an understanding of these systems requires a different conceptual framework. Thus, in order to understand complex systems, we must understand them through a reverse engineering perspective rather than a reductionist perspective. One approach to gaining this understanding is through the use of network representations of living systems.

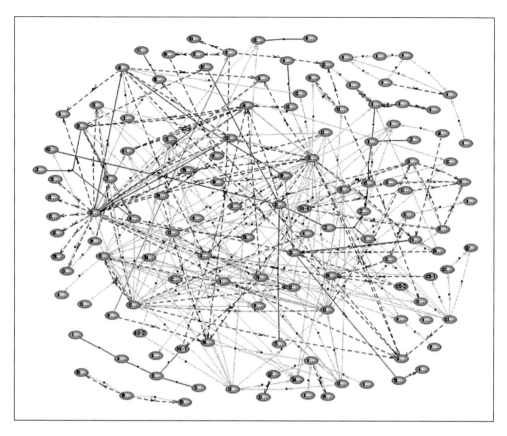

Fig. 1. Illustration of a sample *C. elegans* longevity gene-protein network. See Witten and Bonchev [2] for more details.

Networks and Graphs

In the previous chapter, we introduced the basics of network theoretic methods as a beginning means to understand the complexity of living systems. In particular, we introduced the concept of a graph G that has nodes n_j (or vertices) and edges E_{ij} (a connection between node n_j and node n_i). We illustrate an undirected longevity gene-protein network in figure 1 [1, 2]. We then introduced the idea of the adjacency matrix A. With this simple set of definitions, we now have some powerful tools with which to investigate the structure of a network and how it might inform us about the biological dynamics of the overall network. We began with the concept of *connectivity*. Consider the network in figure 1. We note that some of the nodes appear to have very many connections while others have but a few. Does this imply anything about the network and its behaviors? What information could we glean from this?

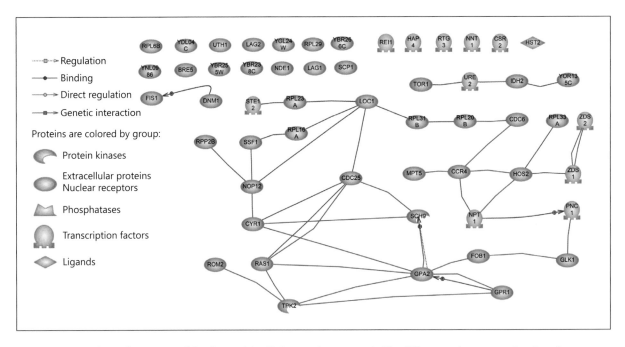

Fig. 2. Illustration of the *S. cerevisiae* RLS extension network. The different color connecting lines indicate different types of interaction; binding (purple), regulation (dotted lines), genetic interaction (green) and direct regulation (grey). The different shapes represent different types of factors involved in the connections.

Creating a Longevity Network: An Example with Yeast

To understand network dynamics, particularly as applied to aging, we first discuss how one can actually go about creating a longevity gene-protein network. The answer is, it is not easy, and it takes a good deal of time. The network in figure 1 took nearly a year of database searching, literature review and peer collaboration to create. Of course, when we started the project, the available databases were not nearly as efficient or sophisticated and the data far less abundant than they are now.

Let us first consider *Saccharomyces cerevisiae*. *S. cerevisiae* was chosen as the model organism for this study because it is well understood, highly studied, and regarded as a good model with which to study aging processes [3]. Yeast has two different ways that it can age; replicative life span (RLS) and chronological life span. When constructing a longevity network, genes having an effect on chronological life span and RLS can both be considered. Chronological life span is a measure of the length of time a nondividing cell can survive, while RLS measures how many times a cell can divide [4, 5]. While both are useful, RLS was chosen because it is easier to study with yeast and has been shown to have an overlap with more complicated organisms [4]. Each of the proteins in the longevity network was shown to increase RLS when removed from the genome as a result of gene knockout studies [6–34]. We illustrate the RLS network for *S. cerevisiae* in figure 2.

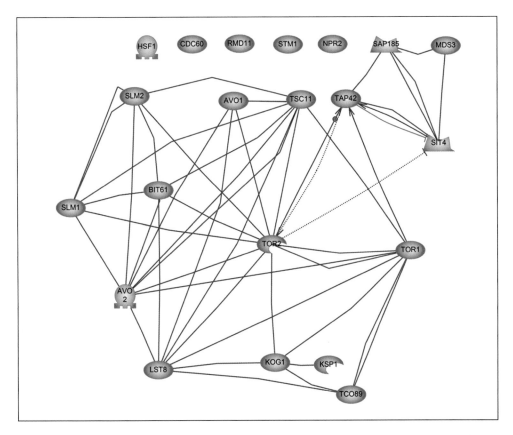

Fig. 3. Illustration of the *S. cerevisiae* TOR network. The different color connecting lines indicate different types of interaction; binding (purple), regulation (dotted lines), genetic interaction (green), expression (blue), protein modification (golden green), promotion of binding (lime green) and direct regulation (grey). The different shapes represent different types of factors involved in the connections.

Notice that there are a number of unconnected nodes. These are our *islands*. They are likely unconnected because we do not know how they are connected in the network.

In this example, we are interested in how different subnetworks (TOR: target of rapamycin, and CRH: cellular response to heat) are tied to the RLS network. And in understanding how the addition of other subnetworks to the RLS network might further inform us about the dynamics of aging at a genetic level. We choose the TOR pathway due to its demonstrated effect on RLS, and we choose the CRH as a model for how the cell responds to environmental stressors [18, 35–37]. Biological aging has been described as a cascading breakdown of processes resulting in decreased ability of an organism to respond to stress, and thus a model of stress response was included [5]. We illustrate the TOR and CRH networks in figure 3 and figure 4. The combined total network TOT is illustrated in figure 5. All of the images were created using Pathway Studio software.

All of the networks were built by mining the literature [for details see 2, 24] and online databases. In particular, we added information from the Saccharomyces Genome

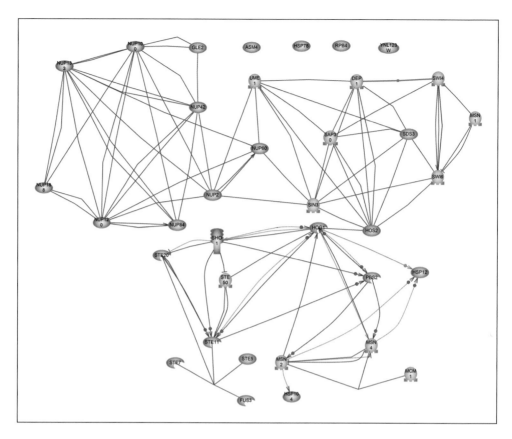

Fig. 4. Illustration of the *S. cerevisiae* CRH network. The different color connecting lines indicate different types of interaction; binding (purple), regulation (dotted lines), genetic interaction (green), expression (blue), protein modification (golden green), promotion of binding (lime green) and direct regulation (grey). The different shapes represent different types of factors involved in the connections.

Database, YEASTRACT, The Comprehensive Yeast Genome Database, The NetAge Database, Sageweb, and AmiGO. Items on the list were verified by comparing them to results from papers that measured the effect on RLS of gene deletions (see previous references). Lists for the TOR as well as the CRH were constructed using gene ontology terms on AmiGO [35] as well as the Saccharomyces Genome Database [37].

Once the lists were prepared, protein-protein interaction data were obtained using a yeast interaction database developed to work with the software program Pathway Studio. A network of direct connections (DC) was constructed as well as a shortest-path network (SP) for the RLS, TOR, and CRH subnetworks. A TOT was also constructed. TOT was constructed by fusing the three networks together into the larger TOT. Not only did Pathway Studio show the connections between proteins, but it also yielded information on their function and the type of interaction. We illustrate a sample list for the CRH network in table 1.

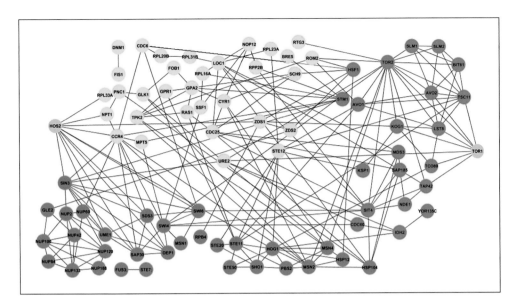

Fig. 5. Illustration of the *S. cerevisiae* combined network TOT. The combined network is composed of RLS, TOR, and CRH. Here, we used the software program cytoscape to graphically represent all three networks combined, where nodes are color coded to show which network the protein it represents is part of. Yellow nodes represent proteins in the RLS network, green for the CRH group, and blue for the TOR network.

Table 1. Connection types for the CRH network

Connection type	n
Binding	39
Direct regulation	17
Expression	2
Genetic interaction	29
Promoter binding	1
Protein modification	2
Regulation	8

The connection types are generated as output from the software. The right-hand column represents the number of each connection type.

Analyzing the Network: An Example with Yeast and *Caenorhabditis elegans*

As we mentioned in the previous chapter, it is natural to conclude that the more edges going in and out of a node, the more likely that the given node is going to be of importance to the network. In order to assist us in understanding the connectivity structure of the network, we create a connectivity plot (fig. 6). To do this, we first count the number of nodes with a given connectivity k where the connectivity varies from zero

Fig. 6. Illustration of a sample connectivity or degree distribution plot for the network in figure 1. See Witten and Bonchev [2] for more details. The rhombs represent the complete distribution. The squares are the data points binned into groups of three. The black solid line is the non-linear regression line. Results are significant at p < 0.05.

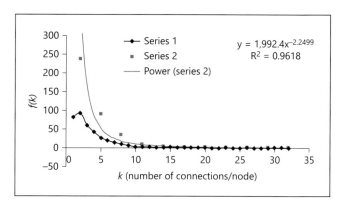

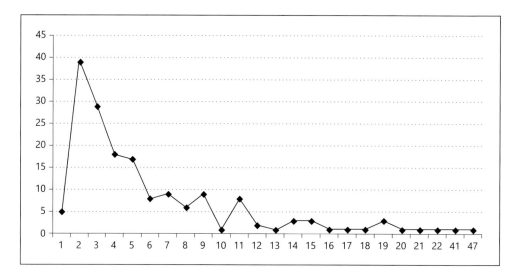

Fig. 7. The degree distribution plot for yeast RLS-SP network. The horizontal axis is the connection number k and the vertical axis is the frequency $f(k)$. Notice that the degree distribution is more irregular and that there are an enormous number of zero and one values in the network. This makes fitting a power curve more difficult and also increases the likelihood that the fit will not be statistically significant due to the small sample size (number of nodes). Sometimes binning can help when there are zero node numbers. However, that also affects the fit.

to the maximum connectivity value. The number of nodes with a given connectivity k is called the frequency of that connectivity and is denoted $f(k)$. Next, plot the frequency $f(k)$ versus the connectivity k. We illustrate this for *Caenorhabditis elegans* in figure 6. Because the *C. elegans* network is large, it has a smoother look to it. Let us look at the yeast RLS shortest-path network (RLS-SP) and construct the power-law graph (fig. 7).

As we mentioned in the previous chapter, studies of the statistical behavior of various network structures have shown that networks can have a small variety of overall

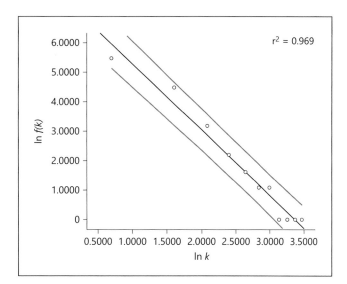

Fig. 8. Illustration of the log-log data and regression curve through the data of figure 6. The outer lines are the 95% confidence interval boundaries for the linear regression estimate. See Witten and Bonchev [2] for more details.

topologies; random, regular, small-world and scale-free. Moreover, many real-world networks can be shown to be small-world or scale-free. Because scale-free networks are ubiquitous and highly relevant to our discussion, let us look at them a bit more closely. How can we determine if we have a scale-free distribution?

Power Plots and Scale-Free Networks

It is hard to interpret a plot like that illustrated in figures 6 and 7. However, we observe that if we take the log of both sides of $f(k) = Bk^{-\gamma}$, the more linear the data plot, the more likely it fits a power curve. This follows because we would have $\ln[f(k)] = -\gamma\ln(k) + \ln(B)$. Thus, networks whose connectivity structure follows a power law of the form $f(k) = Bk^{-\gamma}$ where B and γ are parameters to be estimated should look like negative slope lines if they are scale free. The simplest way to estimate the parameters is to perform a linear regression on the log-log transformed $f(k)$ versus k data, dropping the $k = 0$ data point because there are no connectivities. We found that $B = 1,992.4$ and $\gamma = 2.2499$ with an $r^2 = 0.969$ (fig. 8). Thus, our *C. elegans* longevity gene-protein network [2] can be said to be a scale-free network.

Categorizing Small-World Networks

Due to the unique nature of scale-free networks, a log-log connectivity plot is enough to let you know if you are dealing with a scale-free network. However, this trick does not work for other network forms. Because many biological systems demonstrate

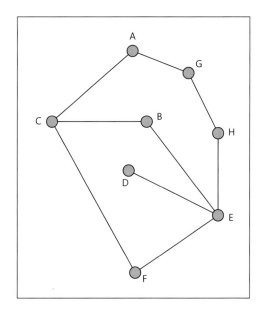

Fig. 9. Illustration of a simple hypothetical network.

Table 2. Sample adjacency matrix A (left block of numbers)

	A	B	C	D	E	F	G	H	A	B	C	D	E	F	G	H
A	0	0	1	0	0	0	1	0	0	0	21	0	23	0	13	0
B	0	0	1	0	1	0	0	0	0	0	29	0	36	0	15	0
C	1	1	0	0	0	1	0	0	21	29	0	15	0	29	0	22
D	0	0	0	0	1	0	0	0	0	0	15	0	21	0	8	0
E	0	1	0	1	0	1	0	1	23	36	0	21	0	36	0	29
F	0	0	1	0	1	0	0	0	0	0	29	0	36	0	15	0
G	1	0	0	0	0	0	0	1	13	15	0	8	0	15	0	14
H	0	0	0	0	1	0	1	0	0	0	22	0	29	0	14	0

On the right, we have multiplied A times itself five times. The non-zero number in the entry represents the number of possible paths between node n_i and node n_j.

small-world network behavior, we briefly examine how to determine whether or not a network is a small-world network.

To help characterize small-world networks, in the previous chapter, we introduced a few new network descriptors. The first was the *average path length* of a network. Path length is the distance or number of edges between two nodes in the network. We used the idea of path length to construct the *minimum path length* between node n_i and node n_j and denoted it by ℓ_{ij}. We now introduce the concept of the *diameter* of a network. The diameter is the largest direct distance between any two nodes in the network. Consider the example network in figure 9.

Now, consider the adjacency matrix A derived from the network in figure 9 and which is illustrated on the left hand side of table 2. The matrix A represents the number of length 1 paths between node n_i and node n_j. If we multiply $A \times A$, the entries that are non-zero represent the number of length 2 paths between each pair of nodes. If we repeat this process exactly $N-1$, then we get the total number of paths of length $1,2,\ldots,N-1$ for the network between each pair of nodes in the network where N is the total number of nodes in the network. The right hand side of table 2 illustrates $A \times A \times A \times A \times A$, which is the number of paths of length five between each pair of nodes. So, for example, there are 29 possible paths of length 5 between node C and node B. At some point between 1 and $N-1$ multiplies, every element in the matrix or its subsequent multiplies A, A^2, A^3, A^4, $\ldots$, A^{N-1} will have been non-zero during the multiply sequence. The diameter is the minimum number of times the adjacency matrix A has to be multiplied by itself so that each entry has taken a value greater than 0 at least once during the multiply sequence. For this particular network, the diameter is 4.

A network is considered a small-world network if the diameter is small relative to the number of nodes in the network. Obviously, 4 is not small relative to 8. To really understand small-worldness, one network sample is not sufficient. You actually need a set of graphs. However, we have only the one network graph. Therefore, we need other means to study the behavior of networks to understand if they are small-world networks. To do this, we introduced the ideas of clustering. Observe that even for a small network such as the one we have illustrated, the calculations can become tedious. We will discuss software at a later point in this chapter.

Node-Node Connectivities

In the previous chapter, we talked about the idea of the *centrality* of a node where centrality is a measure of the 'position' or relative importance of a node in a network. In the literature, there are four main measures of centrality of a node: *degree centrality*, *betweenness centrality*, *closeness centrality* and *eigenvector centrality*. From an aging-related perspective, understanding node centrality of the nodes in a network could lead to potential targets for pharmaceuticals that might help hinder disease progression or extend life span. Briefly, the main centrality measures are:

- *Degree centrality* of a node is denoted by $C_D(n_i)$; it measures the chance that a given node n_i in the network will receive something flowing along the network.
- *Closeness centrality*, denoted $C_C(n_i)$ can be thought of as a measure of how long it will take to send a chemical or other biological signal out from n_i to all of the other nodes in the network.
- *Betweenness centrality*, denoted $C_B(n_i)$ looks at how often, in a network, a given node n_i acts as a bridge along the shortest path between two other nodes. From a biological perspective, knocking out a node with high betweenness centrality would force a signal to reroute itself along a path that was not the shortest path.
- *Eigenvector centrality*, denoted $C_E(n_i)$ is a measure of the 'influence' of a node in a network.

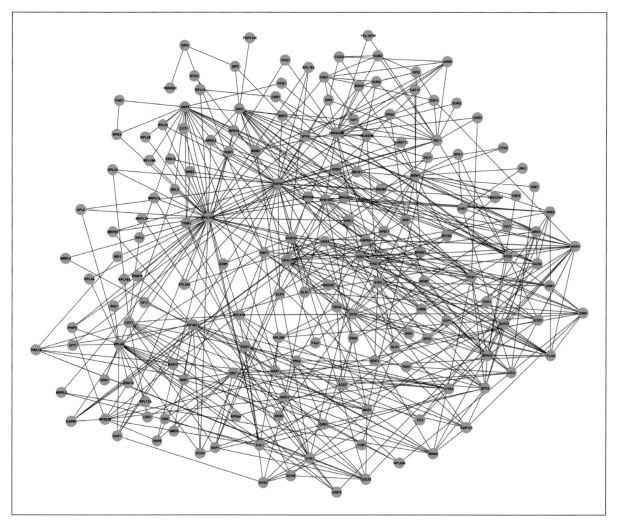

Fig. 10. Illustration of the RLS-SP network. Note how highly connected the one node in the upper left hand side is. That node, by the way, is RPL16B ribosomal 60S subunit protein L16B; N-terminally acetylated, binds 5.8 S rRNA; transcriptionally regulated by Rap1p; homologous to mammalian ribosomal protein L13A and bacterial L13 [37].

Eigenvalue centrality is commonly used as a centrality measure because it is an influence measure for a node, and these could be potential targets for further study or drug design. In figure 10, we illustrate the SP for the RLS network. For a discussion of SPs in aging, see Managbanag et al. [24]. In figure 11, we illustrate the corresponding eigenvalue centrality measures for the various nodes in the RLS-SP network. The larger the eigenvalue centrality, the larger the influence in the RLS-SP network.

Thus, from figure 11, we would infer that *GPA2, CDC25, SCH9* and *CYR1* are influential nodes and therefore likely candidates to investigate further. SCH9 – AGC

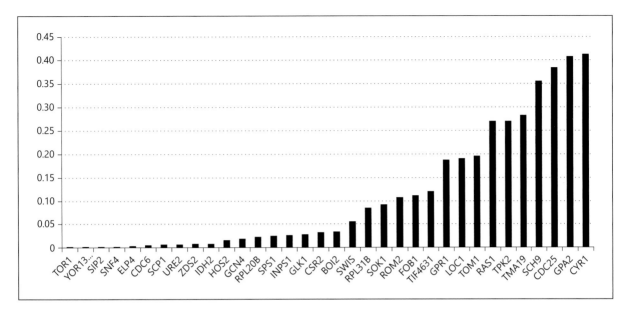

Fig. 11. Illustration of the eigenvalue centrality measures for the yeast replicative shortest-path life span network. The horizontal axis is the node name in the network, the vertical axis is the eigenvalue centrality value. The larger the eigenvalue centrality, the larger the influence in the RLS-SP network.

family protein kinase; functional ortholog of mammalian S6 kinase; phosphorylated by Tor1p and required for TORC1-mediated regulation of ribosome biogenesis, translation initiation, and entry into G0 phase; involved in transactivation of osmostress-responsive genes; regulates G1 progression, cAPK activity and nitrogen activation of the FGM pathway [37]. CYR1 – adenylate cyclase is required for cAMP production and cAMP-dependent protein kinase signaling; the cAMP pathway controls a variety of cellular processes, including metabolism, cell cycle, stress response, stationary phase, and sporulation [37]. CDC25 – Membrane-bound guanine nucleotide exchange factor; indirectly regulates adenylate cyclase through activation of Ras1p and Ras2p by stimulating the exchange of GDP for GTP; required for progression through G1 [37]. GPA2 – Nucleotide-binding α-subunit of the heterotrimeric G protein interacts with the receptor Gpr1p, has signaling role in response to nutrients [37]. Witten and Bonchev [2] illustrate these concepts for the *C. elegans* longevity gene network illustrated in figure 1 of that paper.

Most network analysis programs calculate the basic centrality and other measures. The algorithms for these calculations are tedious and not trivial to program. Therefore, it is better to use one of the programs discussed in the upcoming software section rather than to write your own programs to make the calculations. In table 3, we illustrate the properties of all four of our original networks and their corresponding SPs.

Table 3. Some of the basic network variables for the original CRH, RLS, TOR and TOT networks and their corresponding shortest-path networks

Network	Number of nodes	Number of edges	Vertex degree range	Node density	Mean Vertex degree	Mean node distance	Network diameter
CRH DC	28	70	1–8	0.18519	5	3.146	7
CRH SP	119	460	1–35	0.06552	7.7311	2.6	5
RLS DC	37	46	1–6	0.06907	4.383	2.7059	10
RLS SP	169	524	1–47	0.03691	6.2012	2.893	6
TOR DC	16	42	1–11	0.35000	5.25	1.942	4
TOR SP	331	1,500	2–110	0.02746	9.0361	2.855	5
TOT DC	76	158	1–11	0.05544	4.1579	5.774	13
TOT SP	558	2,467	1–110	0.01587	8.853	3.171	6

Fragmentation (DC)	0.59158
Fragmentation (SP)	0.51203
Clustering coefficient (DC)	0.20142
Clustering coefficient (SP)	0.04330

Interpreting the Results

From the eigenvalue centrality, we have already seen a number of genes worth investigating due to their influential nature in the networks. From the graph of the DC, we discovered that the TOR1 protein was shared between the RLS and TOR networks, and the HOS2 gene was shared between the RLS and the CRH networks. It was also discovered that the TOR and CRH networks were densely connected to the RLS network. However, there were relatively few connections between the CRH and TOR networks. TOR1 is responsible for PIK-related protein kinase and rapamycin target; subunit of TORC1, a complex that controls growth in response to nutrients by regulating translation, transcription, ribosome biogenesis, nutrient transport and autophagy; involved in meiosis [37]. HOS2 is histone deacetylase and subunit of Set3 and Rpd3L complexes; required for gene activation via specific deacetylation of lysines in H3 and H4 histone tails; subunit of the Set3 complex, a meiotic-specific repressor of sporulation-specific genes that contains deacetylase activity [37]. We observe that all of these targets are related to growth and division in some way.

We were able to demonstrate that the SPs followed a power-law distribution. This was not the case in the DC perhaps due to their relatively small sample size. Mean vertex degree was noticeably different between the SPs and DCs. It was more pronounced in the TOR network, the shortest-path mean vertex degree being nearly double what it was in the DCs. This was even more noticeable between the TOTs, which was more than double. There was a large difference in node densities with the SPs having ones lower than the DCs. Between the RLS networks, this was far less pronounced with the SP having half the node density of the DC. The TOR shortest-path node density was a one tenth of the DCs. The network diameter was similar for the TOR DCs and the SPs, yet for the RLS and TOTs the shortest-path diameter was half what it was in the DC.

Software for Network Analysis

Due to the fact that many networks have large numbers of nodes and connections, it is not possible to hand-calculate the various network descriptors that we have discussed. Over the past decade, a number of network analysis software packages have become available. Two of the most commonly used packages are Pajek, available at http://vlado.fmf.uni-lj.si/pub/networks/pajek/, and Cytoscape, which is available at http://www.cytoscape.org/. Another excellent package is NetworkX from Los Alamos National Laboratories. It can be downloaded at http://networkx.lanl.gov/index.html. All of these packages offer free downloads on numerous computation platforms and operating systems.

One of the challenges in understanding large complex networks, including biological networks, is visualizing them. Both Pajek and Cytoscape offer network visualization tools. However, a number of other visualization tools are now available, and these are very powerful visualization software packages. CFinder, available at http://cfinder.org/ is a cluster and community software package designed for finding and visualizing dense groups of nodes in networks. Gephi, available at https://gephi.org, is an open graph visualization program that allows the user to perform exploratory data analysis on a given network, link analysis and generate high-quality printable network images. There are many other social network analysis software packages now available. The packages frequently allow the user to analyze biological networks as well as other network forms. An excellent discussion of available network analysis and visualization software may be found at http://en.wikipedia.org/wiki/Social_network_analysis_software.

Future Directions

Future directions for the research include adding additional yeast subnetworks that are believed to have a tie to aging processes. In addition, we will add networks that are believed to be unrelated to replicative aging processes. These unrelated networks will serve as control networks. For the TOT of direct interactions, proteins were labeled to show which group they belonged to (TOR, heat-shock, RLS, or shared). It would be helpful to do the same for the TOT of shortest-path connections.

We will also take what we have learned about studying yeast networks, and use this to study protein-protein interaction networks in other species, such as *C. elegans*, *Drosophila melanogaster* and eventually in humans. By using *C. elegans*, a wider variety of genes that have an effect on aging can be studied, i.e. genes such as the FOXO gene.

However, because it is multicellular, the *C. elegans* genome would be more complex, having about 20,000 genes as opposed to only 6,000 in yeast. Homologs to yeast genes/proteins in other organisms can be investigated as possible important genes/proteins. Using human interaction networks allows the study to be directly related to the study of aging in humans, which is the ultimate goal. With an expanded yeast network, it will be easier to show links between existing data and studies of other model organisms. It might also help guide decisions on which networks to study in *C. elegans* and humans.

Closing Thoughts

In the previous sections, we introduced a large number of concepts and constructs that are based upon the premise that biological systems can be represented as network graphs. These concepts described how network nodes were interconnected and the consequences of certain specific classes of connectivity and network structure. At the 1982 Palo Alto American Mathematical Society meeting, Witten presented a paper on representing aging using the model of network decay. Of course, in those days, network analysis was not what it is today, and we had next to nothing of the genomic and network level data that we now have. However, even then, it was natural to consider aging as the temporal decay of a hypothetical organismal 'aging network'. How then may we extend these ideas to the study of aging?

While little is currently known about how aging-related networks evolve across the organism's life span, it is reasonable to assume that two possible changes can occur; inactivation of active nodes/activation of inactive nodes and loss of connectivity/increase in connectivity. How or why nodes become inactive or edges disappear is irrelevant here; just that they do. It turns out that the structure of small-world networks, due to their hub connectivity, makes them vulnerable to targeted attacks aimed at specific hubs. Attacks that knock out essential genes are knocking out the life span network because the organism dies when an essential gene is knocked out. Thus, essential genes are critical hub genes [2, 38, 39]. Small-world neural networks have been shown to exhibit short-term memory capability. This suggests that memory decay, such as that seen in Alzheimer's disease may be related to decay of brain neural network structure in such a way as to remove the small-worldness property of the memory network. Understanding patterns in network decomposition could lead to potential early AD detection and to potential pharmaceutical intervention at earlier points in the disease course.

Connectivity gain and loss also have implications when it comes to discussing the hierarchical modularity of aging-related network architectures. Loss of connectivity through inactivity of a node or through loss of an edge could unlink an entire module of importance. Thus, nodes that connect modules within a larger network are critical to the functioning of the network. Questions around the role of evolutionary processes in the development of network architectures of various organisms may be of importance in understanding how network architectures related to aging processes are constructed.

Why are some components of a network redundant while others are not (see also all of the citations on reliability theory)? What is the role of backup subnetworks? What is the importance of robustness and resilience? Why are some networks more robust to attack [46–50], less fragile than others or more frail [40–44]? How do we balance the need to adapt and evolve with robustness [45]? What, if any, is the association of life span with network architecture? These and many other questions remain to be answered.

Acknowledgements

The authors would like to thank many individuals for their respective support and collaborative kindnesses. In alphabetical order I would like to acknowledge my colleagues and friends: Danail Bonchev, S. Michal Jazwinski, Tom Johnson, Matt Kaeberlein and Brian Kennedy for their support and access to data and software. An expanded bibliography for both chapters is available at http://www.people.vcu.edu/~tmwitten.

References

1 Witten TM: (M,R)-systems, (P,M,C)-nets, hierarchical decay and biological aging: reminiscences of Robert Rosen. Chem Biodivers 2007;4:2332–2344.

2 Witten TM, Bonchev DG: Predicting aging/longevity-related genes in the nematode *C. elegans*. Chem Biodivers 2007;4:2639–2655.

3 Breitenbach M, Jazwinski SM, Laun P: Aging research in yeast. Springer Cell Cycle 2012;10:1385–1396.

4 Longo VD, Shadel GS, Kaeberlein M, Kennedy B: Replicative and chronological aging in *Saccharomyces cerevisiae*. Cell Metab 2012;16:18–31.

5 Steffen KK, Kennedy BK, Kaeberlein M: Measuring replicative life span in the budding yeast. J Vis Exp 2009;28:1209.

6 Anderson RM, Bitterman KJ, Wood JG, Medvedik O, Sinclair DA: Nicotinamide and PNC1 govern lifespan extension by calorie restriction in *Saccharomyces cerevisiae*. Nature 2003;423:181–185.

7 The Basic Biology of Aging. http://www.uwaging.org/genesdb/.

8 Burtner CR, Murakami CJ, Olsen B, Kennedy BK, Kaeberlein M: A genomic analysis of chronological longevity factors in budding yeast. Cell Cycle 2011;10:1385–1396.

9 Defossez PA, Prusty R, Kaeberlein M, Lin SJ, Ferrigno P, et al: Elimination of replication block protein Fob1 extends the life span of yeast mother cells. Mol Cell 1999;3:447–455.

10 Delaney J, Murakami CJ, Olsen B, Kennedy BK, Kaeberlein M: Quantitative evidence for early life fitness defects from 32 longevity-associated alleles in yeast. Cell Cycle 2011;10:156–165.

11 D'Mello NP, Childress AM, Franklin DS, Kale SP, Pinswasdi C, et al: Cloning and characterization of LAG1, a longevity-assurance gene in yeast. J Biol Chem 1994;269:15451–15459.

12 Fabrizio P, Pozza F, Pletcher SD, Gendron CM, Longo VD: Regulation of longevity and stress resistance by Sch9 in yeast. Science 2001;292:288–290.

13 Fabrizio P, Liou LL, Moy VN, Diaspro A, Valentine J, et al: SOD2 functions downstream of Sch9 to extend longevity in yeast. Genetics 2003;163:35–46.

14 Gourlay CW, Carpp LN, Timpson P, Winder SJ, Ayscough KR: A role for the actin cytoskeleton in cell death and aging in yeast. J Cell Biol 2004;164:803–809.

15 Hoopes LL, Budd M, Choe W, Weitao T, Campbell JL: Mutations in DNA replication genes reduce yeast life span. Mol Cell Biol 2002;22:4136–4146.

16 Institute of bioinformatics and Systems biology. http://www.helmholtz-muenchen.de/en/ibis.

17 Kaeberlein M, McVey M, Guarente L: The SIR2/3/4 complex and SIR2 alone promote longevity in *Saccharomyces cerevisiae* by two different mechanisms. Genes Dev 1999;13:2570–2580.

18 Kaeberlein M, Kirkland KT, Fields S, Kennedy BK: Genes determining yeast replicative life span in a long-lived genetic background. Mech Ageing Dev 2005;126:491–504.

19 Kennedy BK, Austriaco NR, Zhang J, Guarente L: Mutation in the silencing gene SIR4 can delay aging in *S. cerevisiae*. Cell 1995;80:485–496.

20 Kim S, Benguria A, Lai CY, Jazwinski SM: Modulation of life-span by histone deacetylase genes in *Saccharomyces cerevisiae*. Mol Biol Cell 1999;10:3125–3136.

21 Kruegel U, Robison B, Dangel T, et al: Elevated proteasome capacity extends replicative lifespan in *Saccharomyces cerevisiae*. PLoS Genet 2011;7:1–16.

22 Lin SJ, Defossez PA, Guarente L: Requirement of NAD and SIR2 for life-span extension by calorie restriction in *Saccharomyces cerevisiae*. Science 2000; 289:2126–2128.

23 Lu JY, Lin YY, Sheu JC, et al: Acetylation of yeast AMPK controls intrinsic aging independently of caloric restriction. Cell 2011;146:968–979.

24 Managbanag JR, Witten TM, Bonchev DG, Fox LA, Tsuchiya M, Kennedy BK, Kaeberlein M: Shortest-path network analysis is a useful approach towards identifying genetic determinants of longevity. PLoS One 2008;3:e3802.

25 Murakami CJ, Burtner CR, Kennedy BK, Kaeberlein M: A method for high-throughput quantitative analysis of yeast chronological life span. J Gerontol Biol Sci Med Sci 2008;63:113–121.

26 Roy N, Runge KW: Two paralogs involved in transcriptional silencing that antagonistically control yeast life span. Curr Biol 2000;1:111–114.

27 Scheckhuber CQ, Erjavec N, Tinazli A, Hamann A, et al: Reducing mitochondrial fission results in increased life span and fitness of two fungal ageing models. Nat Cell Biol 2007;9:99–105.

28 Sinclair DA, Mills K, Guarente L: Accelerated aging and nucleolar fragmentation in yeast sgs1 mutants. Science 1997;277:1313–1316.

29 Sinclair DA, Guarente L: Extrachromosomal rDNA circles – a cause of aging in yeast. Cell 1997;91:1033–1042.

30 Smith DL, McClure JM, Matecic M, Smith JS: Calorie restriction extends the chronological lifespan of *Saccharomyces cerevisiae* independently of the Sirtuins. Aging Cell 2007;6:649–662.

31 Smith ED, Tsuchiya M, Fox L, et al: Quantitative evidence for conserved longevity pathways between divergent eukaryotic species. Gene Res 2008;18:564–570.

32 Steffen KK, MacKay VL, Kerr EO, Tsuchiya M, et al: Yeast life span extension by depletion of 60s ribosomal subunits is mediated by Gcn4. Cell 2008;133: 292–302.

33 Sun J, Kale SP, Childress AM, Pinswasdi C, Jazwinski SM: Divergent roles of RAS1 and RAS2 in yeast longevity. J Biol Chem 1994;269:18638–18645.

34 Tsuchiya M, Dang N, Kerr EO, Hu D, et al: Sirtuin-independent effects of nicotinamide on lifespan extension from calorie restriction in yeast. Aging Cell 2006;5:505–514.

35 AmiGO: http://amigo.geneontology.org/cgi-bin/amigo/go.cgi.

36 Kaeberlein M, Powers RW, Steffen KK, Westman EA, Hu D, et al: Regulation of yeast replicative life span by TOR and Sch9 in response to nutrients. Science 2005;310:1193–1196.

37 SGD, The Saccharomyces Genome Database. http://www.yeastgenome.org/.

38 Witten TM: Reliability theoretic methods and aging: critical elements, hierarchies, and longevity – interpreting survival curves; in Woodhead A, Blackett A, Setlow R (eds): The Molecular Biology of Aging. New York, Plenum Press, 1985a.

39 Witten TM: A return to time, cells, systems and aging. III. Critical elements, hierarchies, and Gompertzian dynamics. Mech Ageing Dev 1985;32:141–177.

40 Agoston V, Csermely P, Pongor S: Multiple, weak hits confuse complex systems. Phys Rev E Stat Nonlin Soft Matter Phys 2005;71:051909.

41 Basset DS, Bullmore E: Small-world brain networks. Neuroscientist 2006;12:512–523.

42 Chan KP, Zhen D, Hui PM: Effects of aging and links removal on epidemic dynamics in scale-free networks. Int J Modern Phys B 2004;18:2534.

43 Csermely P: Strong links are important but weak links stabilize them. Trends Biochem Sci 2004;29: 331–334.

44 Csermely P: Creative elements: network-based predictions of active centres in proteins and cellular and social networks. Trends Biochem Sci 2008;33:569–576.

45 Gallos LK, Makse HA, Sigman M: A small world of weak ties provides optimal global integration of self-similar modules in functional brain networks. Proc Natl Acad Sci USA 2012;109:2825–2830.

46 Gavrilov LA, Gavrilova NS: Models of systems failure in aging; in Conn PM (ed): Handbook of Models for Human Aging. Burlington, Elsevier Academic Press, 2006, pp 45–68.

47 Kriete A: Robustness and aging – a systems-level perspective. Biosystems 2013;112:37–48.

48 Lemke N, Heredia F, Barcellos CK, Dos Reis AN, Mombach JC: Essentiality and damage in metabolic networks. Bioinformatics 2004;20:115–119.

49 Saavedra S, Reed-Tsochas F, Uzzi B: Asymmetric disassembly and robustness in declining networks. Proc Natl Acad Sci USA 2008;105:16466–16471.

50 Huang X, Gao J, Buldyrev SV, Havlin S, Stanley HD: Robustness of interdependent networks under targeted attack. 2010. http://arxiv.org/abs/1010.5829v1.

Tarynn M. Witten
Center for the Study of Biological Complexity, Virginia Commonwealth University
PO Box 842030, 1000 West Cary Street
Richmond, VA 23284 (USA)
E-Mail tmwitten@vcu.edu

Yashin AI, Jazwinski SM (eds): Aging and Health – A Systems Biology Perspective.
Interdiscipl Top Gerontol. Basel, Karger, 2015, vol 40, pp 35–48 (DOI: 10.1159/000364928)

Computational Systems Biology for Aging Research

Mark T. Mc Auley[a] · Kathleen M. Mooney[b]

[a]Faculty of Science and Engineering, Thornton Science Park, University of Chester, Chester, and
[b]Faculty of Health and Social Care, Edge Hill University, Ormskirk, UK

Abstract

Computational modelling is a key component of systems biology and integrates with the other techniques discussed thus far in this book by utilizing a myriad of data that are being generated to quantitatively represent and simulate biological systems. This chapter will describe what computational modelling involves; the rationale for using it, and the appropriateness of modelling for investigating the aging process. How a model is assembled and the different theoretical frameworks that can be used to build a model are also discussed. In addition, the chapter will describe several models which demonstrate the effectiveness of each computational approach for investigating the constituents of a healthy aging trajectory. Specifically, a number of models will be showcased which focus on the complex age-related disorders associated with unhealthy aging. To conclude, we discuss the future applications of computational systems modelling to aging research.   © 2015 S. Karger AG, Basel

Aging has intrigued and troubled scholars since the beginning of civilization. It is a process that can be described generally as the changes that take place during the life span of an organism which progressively renders them more likely to die. The alterations that bring about a gradual increase in the probability of mortality involve all aspects of biology, from molecular mechanisms to whole-body physiological systems. Moreover, there is little doubt that aging is modulated extrinsically by diet, while intrinsically the velocity of aging also appears to be shaped by a wide variety of genetic mutations. For instance, mutations to daf-2/daf-16 regulate life span in the nematode [1], while the FOXO3A genotype has been strongly linked with variations in human longevity [2]. Paradoxically, genetic homogeneity does not mean the velocity of aging will be the same, as genetically identical species can display a variety of aging rates [3]. Furthermore, evolution has given rise to significant life span variations between different species [4]. Aging is also seen as central to the understanding of many disease states; for example, in certain tissues the accumulation of senescent cells can lead to cancer via a pro-inflammatory response [5], while neurodegeneration underpins the progression of Alzheimer's (AD) and Parkinson's disease [6]. Moreover, free radical damage has been implicated in a variety of disease pathologies from cardiovascular disease (CVD) to dementia. Historically, biologists have investigated the complexities of aging using conventional wet laboratory techniques; however, it is increasingly recognized that to fully appreciate the uniqueness of aging, systems biology approaches are a necessity [7]. A fundamental aspect of systems biology is computational systems modelling, a procedure which involves the development of in silico models. Such models are ideal for describing the innate complexity and dynamics of aging. However, it is often misunderstood as to what exactly computational systems modelling is. It is not statistical data analysis, the three-dimensional visualization of proteins or database mining; instead, it involves using a computer to quantitatively represent the components of a biological system of interest. How the components interact based on current biological understanding is described with mathematical equations. The computer then simulates the interactions between the components to give an overall graphical account of the dynamics of the system [8]. Thus, computational systems modelling can be easily integrated with other disciplines under the systems biology umbrella, as quantitative data from diverse fields including genomics, metabolomics and proteomics can be utilized to inform model construction and refinement. Moreover, model predictions can be used to direct the future design of wet laboratory experiments and also give insights into how a biological system will behave under a wide-variety of different conditions. For instance, the proposed effects of the aging process can be incorporated into a model by including something as straightforward as the age-associated decline in the activity of the key enzymes of the cellular pathway of interest. Despite the clear advantages outlined above, the utility of modelling to aging research can often be overlooked, or traditional gerontologists can be sceptical about the validity of the model or the modelling process generally. Thus, it is important to extend further the rationale for using computational systems modelling and why it is central to improving our understanding of the aging process.

Rationale for Using Systems Modelling for Aging Research

As outlined, computational models are capable of the quantitative representation and analysis of biological systems, something that is not always possible to achieve in a wet laboratory for a number of reasons. Firstly, biological systems are both inherently detailed and inherently complex. This level of detail and complexity gives rise to a diverse web of overlapping metabolic networks which are comprised of multiple connections between each node in the network. Many of the nodes interact in a non-linear fashion and often communicate with each other via sophisticated feedback or feed-forward loops. This places a significant cognitive burden on the human brain to retain this level of complexity and detail. For instance, if the activity of NAD+ dependent deacetylases, commonly referred to as sirtuins are explored, such complexity becomes apparent as the seven mammalian sirtuins perform numerous interrelating actions and modulate a number of pathways connected to age-related disease [9]. Likewise, the mammalian target of rapamycin (mTOR) pathway is equally complex. This system is known to regulate life span in model organisms, and recently has been suggested as a central intracellular regulator, mechanistically connecting aging, oxidative stress and cardiovascular health [10]. Thus, it is highly improbable that one can reason about such complex systems by human intuition alone and as such computational modelling offers a complimentary means of dealing with the complexity associated with aging. Another reason for using the systems approach is to identify and unravel molecular and biochemical hubs that are key regulators, whose robust dynamics ultimately impact the health of tissues and whole-organ systems. To this end, computational systems biology is beginning to accommodate the representation of biological systems in a multi-scale way [11, 12]. This type of representation contrasts with many conventional methodologies which focus on a small manageable component of a biological system. This is particularly significant for aging, as the most probable way to gain a deeper understanding of this intriguing phenomenon is to investigate the synergistic behaviour of cells, tissues and organ systems. The next section will explore further the advantages of computational systems biology compared to conventional approaches to studying aging.

Advantages over and Interactions with Conventional Techniques

There are many conventional approaches that can be used to study aging. These experimental methodologies have been valuable in aiding our understanding of the aging process and will have a role to play in future aging research; however, such methodologies have limitations. If for example longitudinal studies are examined, this approach certainly has value; however, it can be resource intensive, expensive and time consuming. Most significantly, this approach will not offer immediate benefits for an aging Western population which urgently requires remedies to diseases such as dementia, which almost half of the oldest old (those ≥85 years) in the USA and UK suffer from [13]. Cross-sectional studies, where individuals of varying ages from a population are assessed at the same time point are not as costly as longitudinal investigations. However, distinguishing cause and effect in cross-sectional studies from straightforward association is inherently difficult. As an example, recent evidence has indicated an association between the decline in global DNA methylation and age in humans. DNA methylation is an epigenetic mark that plays an important role in gene expression, gene imprinting and transposon silencing. Paradoxically, advancing age has been associated with the hypermethylation of certain genes, which can result in age-related disease [14]. If a cross-sectional study was conducted to examine DNA methylation status in a cohort of individuals, this phenomenon would more than likely be apparent. However, it would be challenging to disentangle its causes, as a wide variety of intrinsic and extrinsic factors are conjectured to modulate DNA methylation. These factors include a methyl-deficient diet, genetic polymorphisms within the folate pathway and age-related alterations to the activity of DNA methyl transferases, the family of enzymes responsible for transferring methyl groups to the DNA molecule [14]. It is possible that heterogeneous individual combinations of these factors could independently result in the methylation paradox and a cross-sectional study would not be able to unravel this. The significance of biological heterogeneity is further emphasized by the knowledge that clonal populations of cells display significant phenotypic variations. This phenomenon is suggested to arise from stochasticity or noise in gene expression [15]. Aging researchers need to be acutely aware of biological stochasticity and that simulations by computational systems models are capable of representing both inter-individual and inter-cellular stochasticity [16]. When studying aging, it is also important to take account of the ethical considerations, for instance dietary intervention studies are regularly employed to explore potential nutrients that could modulate the aging trajectory; however, there is a moral imperative to consider here. For example, rodents are routinely used to investigate dietary regimes in aging research; but it could be argued that it is unethical to overuse animals in studies of this nature. Significantly though, model organisms have helped to reveal that caloric restriction (CR) can extend life span [17]. However, this raises the issue of whether such findings can be translated to humans, as many difficulties surround these investigations, not least that extended timeframes are needed to decipher the optimal regime most beneficial to healthy aging. It is also important to be cautious when making inferences about the potential effects of CR in humans. We need only look to the disciplines of toxicology and pharmacology to recognize that the physiology of animals does not always translate well to humans. Thus, an in silico human representation of CR would be worth establishing prior to any trial of CR in humans, as modelling could help to reveal any potential dangers of this regime. This is not improbable as computational systems models are currently used to study the long-term effects of diet on the pathological signatures that characterize unhealthy aging [18]. Thus, computational systems modelling can overcome a variety of challenges by providing a framework for aging-centred questions that are unsuitable to test with conventional approaches.

Computational Systems Modelling Approaches for Aging Research

Modelling approaches differ significantly from traditional in vivo or in vitro techniques used to study aging. Firstly, a model can be used as a cheap and rapid test bed for hypothesis exploration. For example, computational models have long been used for testing life history theories that attempt to frame aging within an evolutionary template [19]. Moreover, no matter what framework is used, constructing a model can improve or augment our understanding of the age-related process under examination. This is a result of having to consider the system of interest in an unambiguous and precise fashion using mathematics; and there are several mathematical frameworks which can be adopted to deal with the complexities of aging. The theoretical framework that is employed will depend on the nature of the system to be modelled. Importantly however the model needs to encapsulate the biological essence underpinning the aging process under consideration, and the framework that is employed should be directly informed by biological evidence and not by modeller bias for a particular approach.

Ordinary Differential Equations and Partial Differential Equations

This approach treats biological systems as reaction networks, which can be represented mathematically by ordinary differential equations (ODEs). ODEs are known as ordinary because they depend on one independent variable (time), and it uses the assumption that biological species exist in a well-mixed compartment, where concentrations can be viewed as continuous. It also assumes that large numbers of molecules are involved in reactions and that the average behaviour of the population of molecules is not influenced by variability [20]. ODEs can be coded on the computer and an algorithm solves them numerically to produce a deterministic output. They are the most common mathematical framework used in computational systems biology; however, they are unsuitable for modelling transport processes, diffusion, molecular spatial heterogeneity and stochasticity. The latter of these limitations is important for aging research as intracellular processes such as oxidative stress are often viewed as stochastic events. Recent examples of ODE models that have been employed in aging research include deterministic models used to represent apoptosis [21], immunosenescence in humans [22], and cardiac ventricular dimension alterations during aging in mice [23]. In contrast to ODEs, partial differential equations (PDEs) are multivariable functions with partial derivatives. Not as ubiquitous as ODE models, the main advantage of PDEs is the ability to handle both spatial and temporal dependencies. This is best demonstrated by a recent model of tumour growth, which included cell age, cell size, and the mutation of cell phenotypes [24]. Moreover, it also incorporated proliferating and quiescent tumour cells indexed by successively mutated cell phenotypes of increasingly proliferative aggressiveness.

The model was able to structure tumour cells by both cell age and cell size. A disadvantage of PDE models is that they can be computationally intensive and thus slow [20].

Stochastic Reaction Networks and Probability-Based Models

Stochastic reaction models attempt to represent the discrete random collisions between individual molecules, which is vital when considering that random accumulation of cellular damage has long been implicated with intracellular aging. This type of reaction is suggested to take place if the molecules exist in small numbers or there are fluctuations in their behaviour, for instance variations in cellular free radical levels. Stochastic simulations treat molecule reactions as random events. Computationally, this approach involves an algorithm treating each reaction in the model as a probability/propensity function, e.g. reactions have different probabilities of occurring, which can be altered based on the reaction type. A stochastic algorithm is not concerned with average behaviour, rather the probabilistic formulation determines firstly when the next reaction occurs and secondly what reaction it will be [16]. Due to its historical connection with the free radical theory of aging, mitochondrial/oxidative stress models are commonplace. Recently, a stochastic systems model was used to simulate mitochondrial function and integrity during aging [24]. The model demonstrated that cycles of fusion/fission and cell degradation are required to maintain optimal levels of mitochondria, even during periods of stochastic damage [25]. Another recent model by Kowald and Kirkwood [26] examined the accumulation of mitochondrial DNA deletions with age in post-mitotic cells. Computer simulations were used to study how different mutation rates affect the extent of heteroplasmy. The model showed that random drift works for life spans of around 100 years, but for short-lived animals, the resulting degree of heteroplasmy was incompatible with experimental observations [26]. Another recent stochastic model focused on the age-related factors that contribute to neurodegeneration by investigating the potential role of glycogen synthase kinase 3 and p53 in AD [27]. The model was able to predict that high levels of DNA damage leads to increased activity of p53 [27]. A model based on the same field of study by Tang et al. [28] illustrates the complementary nature of computational modelling and wet laboratory experimentation. The authors used fluorescent reporter systems imaged in living cells and computer modelling to explore the relationships of polyQ, p38MAPK activation, generation of reactive oxygen species, proteasome inhibition and inclusion body formation. Several other probability/stochastic network models have attempted to replicate the dynamics of telomere erosion. For instance, a computational model was able to explore the idea that telomere uncapping is the main trigger for cellular senescence [29]. A more recent stochastic model made the assumption that cell division in each time interval is a random process whose probability decreases linearly with telomere shortening [30]. Computer simulations of this model were also able to provide a qualitative account of the growth of cultured human mesenchymal stem cells [30]. Variability in biological systems can also be represented with a bayesian network (BN). BNs are a type of probabilistic network graph, where each node within the graph represents a variable. Nodes can be discrete or continuous and are connected to a probability density function, which is dependent on the values of the inputs to the nodes [31]. Recently, a special type of BN called a dynamic BN was applied to the Baltimore Longitudinal Study of Aging. The advantage of this approach over conventional BNs was its ability to model feedback loops. The model showed that interactions among regional volume change rates for a mild cognitive impairment group were different from that of a 'normal' aging cohort [32]. A limitation of BNs is that they are entirely probabilistic and do not represent biological systems mechanistically.

Petri Net Models, Boolean Networks and Agent-Based Models

Petri nets are a directed bipartite graph, with two types of nodes, called places and transitions, which are represented diagrammatically by circles and rectangles, respectively. Circles represent 'places' while rectangles represent 'transitions'. Places and transitions are connected via arrows/arcs. Each circle or place contains a number of tokens which is a kin to a discrete number of biochemical molecules, while the stoichiometry is indicated by the weight above the arrow/arc. Tokens can be both consumed and produced within the Petri net, while a Petri net functions by input-output firing at the 'transitions' within the network. The 'firing' of transitions is a kin to a biochemical reaction taking place. The firing of 'transitions' is controlled incrementally using time steps. There are many different variants of Petri net, including coloured, hybrid, continuous and stochastic, each having a slightly different mode of operation. Petri nets are ubiquitously employed to study genetic regulatory networks [33]. From an aging perspective, a recent Petri net model involved modelling the high osmolarity glycerol signalling pathway, an important regulator of several transcription factors that respond to oxidative stress [34]. The model focused on *Saccharomyces cerevisiae* and was able to successfully integrate key signalling, metabolic and regulatory processes in a systems orientated fashion. Boolean network models are also comprised of nodes that can either be in an 'on' or 'off' state. The dynamics of the model are acted out by a series of time steps, with the state of each Boolean variable being updated at each time step. Similar to Petri nets, Boolean models are regularly employed to examine gene regulatory networks. A recent example of a Boolean model relevant to aging research described the behaviour of the apoptosis network. The model provided insights into the interactions between pro- and anti-apoptotic factors [35]. Agent-based models have been increasingly used in aging research also [36]. This is a rule-based approach which is used to investigate biological systems using clusters of independent agents whose behaviour is underpinned by simple rules. These agents are capable of interacting with one another through space and time. Agent-based models have been applied to many areas of aging research, including signalling pathways, and immune responses. An agent-based model has recently been used to model the NF-κB (nuclear factor-κ light chain enhancer of activated B cells). The model incorporated individual molecules, receptors, genes and structural components such as actin filaments and cytoskeleton, while providing a detailed outline of this network [37].

Model Building

The steps in model building in aging research are presented in figure 1.

Step 1: Selecting a System to Model, and Step 2: Checking for Previous Models

Increasingly, modellers are becoming part of the infrastructure of modern wet laboratories, and in theory computational modelling should directly compliment the other systems biology techniques outlined in this book thus far. Therefore, the direction the computational model takes should be informed by the overall research focus of the wet laboratory and should also be integrated with other laboratory experiments [38]. Once an aging-focused system is identified, it is necessary to determine whether the model will simply describe the systems of interest or whether it will focus on predicting the behaviour of the system (a hypothesis-driven model). This decision should be determined by the goals and motivations of the research team. The team will then be required to decide on the components of the model. This is an abstract process, and it is not possible to include every biological species or reaction. As a rule of thumb, model boundary points should be informed by the idea or hypothesis that is under consideration. It is also important to perform a literature search to determine if the system of interest has been modelled previously. This step can be facilitated by the BioModels database, an archive of published peer-reviewed systems models (http://www.ebi.ac.uk/biomodels-main). Models archived in the BioModels database are coded in the model exchange framework, the Systems Biology Markup Language (SBML; http://sbml.org/Main_Page). If no suitable model exists, it will be necessary to develop a list of biological species and to determine how they interact with each other before visually displaying their interactions in a network diagram.

Step 3: Network Diagram Construction, and Step 4: Deciding on a Mathematical Framework

A network diagram is necessary to outline precisely how the biological species interact and to illustrate model boundary points. A variety of approaches can be used to do this,

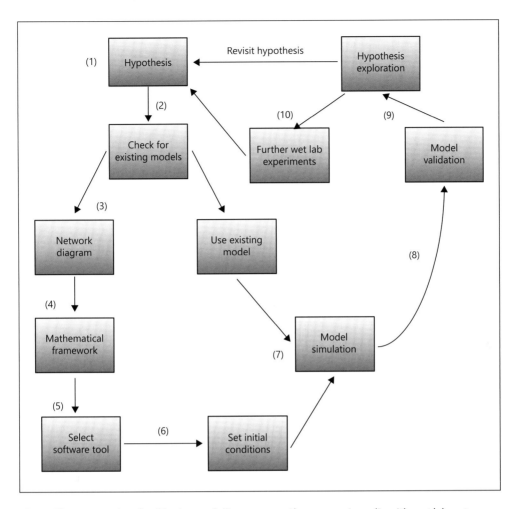

Fig. 1. The 10 steps involved in the modelling process; the process is cyclic with wet laboratory experimentation generating a hypothesis which in turn can be tested by constructing a model; the model in turn feeds further wet laboratory experimentation.

and recently an attempt has been made to standardize how network diagrams are represented using a framework called Systems Biology Graphical Notation [39] which could become the standard means of representing models diagrammatically in the future. To illustrate the network building process, an example of an elementary model of the mTOR signal cascade was developed (fig. 2). The purpose of including this diagram was firstly to illustrate the precise nature of network diagrams. Secondly, the diagram emphasizes that one must abstract when model building. For example, the mTOR signal cascade is a complex network, with >50 components; thus, it was necessary to be selective in order to identify key hubs in the pathway. The network diagram (hypothetical model) commences with the extrinsic stimulation of P13K by growth factors such as those from the

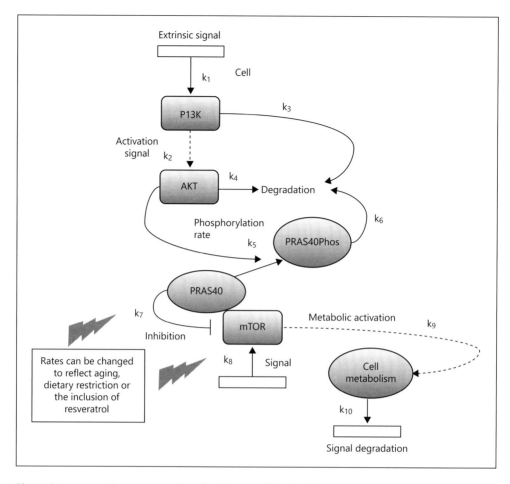

Fig. 2. Diagrammatic representation of mTOR signalling for illustrative purposes. Arrows represent stimulation or conversion reactions; while feedback inhibition is represented by T-shaped arrows. k_1–k_{10} represent kinetic reaction rate constants for each of the steps in the model. k_1 = Rate of activation of P13K; k_2 = rate of activation of AKT; k_3 = degradation rate of P13K; k_4 = degradation rate of AKT; k_5 = rate of PRAS40 phosphorylation; k_6 = rate of PRAS40 degradation; k_7 = inhibition of mTOR signal; k_8 = mTOR signal input; k_9 = metabolic activation; k_{10} = metabolic signal degradation; P13K = phosphoinositide 3-kinase; AKT = protein kinase B; PRAS40 = proline-rich AKT1 substrate 1; PRAS40Phos = phosphorylated PRAS40.

insulin-like growth factor family. This is significant from an aging perspective as CR inhibits the activation of this pathway. AKT is activated by P13K in a manner which depends on the rate at which P13K has been activated (reactions are indicated by arrows, with their kinetic reaction rates indicated by the symbols k_1–k_{10}). Both P13K and AKT have degradation rates. From an ageing perspective it would be worthwhile investigating how changes to these rates impact the system as a whole. Alterations to the levels of AKT have been implicated in the progression of age-related diseases such as cancer and type 2 diabetes; therefore, this is another aspect of the model that could be explored from an intrinsic aging perspective. Dietary regimes associated with longevity could also be investigated; for example, it would be straightforward to incorporate the effects of CR on this pathway or to include the proposed inhibitory effects of the phenolic compound resveratrol.

Step 5: Identify a Suitable Modelling Tool, and Step 6: Setting the Initial Conditions/ Parameters

There are many software tools available to build models of biological systems. Examples include commercial software packages such as Mathematica and MATLAB, while non-commercial tools include Copasi (http://www.copasi.org), CellDesigner (http://www.celldesigner.org/) and PyCml (https://chaste.cs.ox.ac.uk/cellml/). Until recently, it was necessary to learn how to programme competently to construct a computational model, which made the discipline inaccessible to many bioscience researchers. Recently, significant progress has been made, and many tools now come with a graphical user interface (GUI), for instance Copasi [40] and CellDesigner have intuitive GUIs. If the model is kinetic based, setting the initial conditions and parameters involves establishing the initial concentrations of the various biological species and giving each rate law a value. There are many online resources which can be utilized to help with this process. For example, BRENDA (http://www.brenda-enzymes.org/) and SABIO-RK (http://sabio.h-its.org/) archive the details of a wide variety of kinetic data including V_{max} and K_{cat} values which can be used to inform model parameterization.

Step 7: Model Simulation, and Step 8: Model Validation/Parameter Inference

The output from a simulation will depend on the type of mathematical framework that underpins the model. For example, a deterministic solution will always have the same output for a given set of initial conditions and parameters. A stochastic simulation will not produce the same output given a set of initial conditions and parameters. Output from the model can be compared with appropriate time course data to compare the dynamics of the system with its biological counterpart. The sensitivity of the model can also be explored by making adjustments to the model parameters/initial species concentrations. If the model does not compare well to the behaviour of the biological system, it will be necessary to 'fine tune' the parameters to ensure the behaviour of the model is consistent with the dynamics of the biological system. Certain software tools are capable of optimizing a parameter set (or sets) which is consistent with the experimental output. For example, the software tool Copasi has a number of inbuilt statistical techniques to facilitate parameter optimization.

Step 9: Hypothesis Examination, and Step 10: Further Wet Lab Experimentation

If the output of the model appears to be a realistic interpretation of the dynamical behaviour of the system, the model can be used as a predictive tool. If the model does not appear to be a realistic interpretation, one can refine the research question/model; thus, model building is a cyclic process that involves continual revalidation and re-evaluation of the model. If satisfied with the model it can be coded in an exchange format and several exist for computational models, including the Cell Markup Language (Cell-ML; http://www.cellml.org/) and SBML (http://sbml.org/Main_Page). Presently, SBML is the leading exchange format in systems biology and has been evolving since 2000 thanks to an international community of software developers and users.

Computational Systems Models of Aging – From Cell to Whole Body

As outlined, the aging process is inherently complex with a multitude of overlapping relationships that communicate over several different levels. This complexity is a direct result of the underlying multi-scale interconnectivity and interplay of a diverse range of molecular, biochemical and physiological processes. There is no doubt that aging and age-related diseases are a manifestation of the dysregulation and dysfunction of these systems. As a result of the multi-scale nature of biological systems, various different levels of abstraction have been used to create models of a diverse array of processes relating to aging. In the main, systems computational models are cellular in nature; however, recently several aging researchers have come to the conclusion that cellular models, although important are an insufficient means of representing the holistic nature of the aging process and its interaction with age-related pathologies. Consequently, several whole-body computational systems models have been developed. It is not possible to discuss every model; therefore, selections have been restricted to those that illustrate eloquently the diversity and utility of whole-body systems models which have been applied to aging research. For instance, a recent whole-body systems model of cholesterol metabolism was used to explore the interaction of this system with intrinsic aging. The model was able to show that changes to intestinal cholesterol absorption due to the aging process could result in a rise in low-density lipoprotein cholesterol (LDL-C), a key pathological signature of CVD. Moreover, the model also showed that decreasing the rate of hepatic clearance of LDL-C from half its initial value by age 65 years can result in the significant elevation of LDL-C [18]. Other age-related whole-body models have focused on brain aging and dementia. For example, a novel whole-body computational model integrated specific brain regions associated with AD together with the physiological regulation of the stress hormone cortisol. The rationale underpinning the model was to investigate the possible role elevated levels of cortisol have in damaging the hippocampus, the brain region which is the core pathological substrate for AD. The model was able to replicate the in vivo aging of the hippocampus. Moreover, both acute and chronic elevations in cortisol increased aging-associated hippocampal atrophy and concomitant loss in the activity of the hippocampus. The model was also used to investigate potential interventions such as physical activity which could be used to mitigate the effects of aging and cortisol damage to the hippocampus [41].

Conclusions

Computational systems modelling is a novel integrated approach that provides a powerful foundation for gaining an in-depth understanding of how human metabolism is perturbed by aging. This chapter has highlighted the rationale for using computational systems models. The steps involved in the model building process were also outlined, and a wide variety of models from cellular to whole body were discussed that emphasized the utility of modelling to aging research. It is highly probable that in future years computation systems modelling will be further embedded within systems biology. This is something that the aging research community will benefit from as coming years offer the possibility of models being connected together to create a holistic picture of the aging process from genes through to whole organ systems. Such models could focus on multi-scale responses to nutrients or physical activity over extended time frames. In order to achieve this goal, there is little doubt innovative collaborations are a necessity. As this chapter has highlighted, building computational models is a highly collaborative effort that requires considerable interaction between several disciplines. Thus, it is not a process that should occur in isolation as it needs to be firmly integrated within the systems biology paradigm. Working together mathematicians, computer scientists and experimental biologists will be able to provide valuable insights into how robust biological systems break down due to the aging process. Such insights will no doubt contribute to the development of strategies which help to prolong healthy life and delay age-related diseases such as CVD and dementia.

References

1 Kenyon CJ: The genetics of ageing. Nature 2010;464: 504–512.

2 Willcox BJ, et al: FOXO3A genotype is strongly associated with human longevity. Proc Natl Acad Sci USA 2008;105:13987–13992.

3 Kirkwood TB, Finch CE: Ageing: the old worm turns more slowly. Nature 2002;419:794–795.

4 Austad SN: Comparative aging and life histories in mammals. Exp Gerontol 1997;32:23–38.

5 Sikora E, Bielak-Zmijewska A, Mosieniak G: Cellular senescence in ageing, age-related disease and longevity. Curr Vasc Pharmacol 2013, Epub ahead of print.

6 Bredesen DE, Rao RV, Mehlen P: Cell death in the nervous system. Nature 2006;443:796–802.

7 Kirkwood TB: Systems biology of ageing and longevity. Philos Trans R Soc Lond B Biol Sci 2011;366:64–70.

8 Kitano H: Computational systems biology. Nature 2002;420:206–210.

9 Nakagawa T, Guarente L: Sirtuins at a glance. J Cell Sci 2011;124:833–838.

10 Johnson SC, Rabinovitch PS, Kaeberlein M: mTOR is a key modulator of ageing and age-related disease. Nature 2013;493:338–345.

11 Krauss M, et al: Integrating cellular metabolism into a multiscale whole-body model. PLoS Comput Biol 2012;8:e1002750.

12 Walpole J, Papin JA, Peirce SM: Multiscale computational models of complex biological systems. Annu Rev Biomed Eng 2013;15:137–154.

13 Ferri CP, et al: Global prevalence of dementia: a Delphi consensus study. Lancet 2005;366:2112–2117.

14 Muers M: Gene expression: disentangling DNA methylation. Nat Rev Genet 2013;14:519.

15 Kaern M, et al: Stochasticity in gene expression: from theories to phenotypes. Nat Rev Genet 2005;6:451–464.

16 Wilkinson DJ: Stochastic modelling for quantitative description of heterogeneous biological systems. Nat Rev Genet 2009;10:122–133.

17 Heilbronn LK, Ravussin E: Calorie restriction and aging: review of the literature and implications for studies in humans. Am J Clin Nutr 2003;78:361–369.

18 Mc Auley MT, et al: A whole-body mathematical model of cholesterol metabolism and its age-associated dysregulation. BMC Syst Biol 2012;6:130.

19 Kirkwood TB: Evolution of ageing. Nature 1977;270:301–304.

20 Chen WW, Niepel M, Sorger PK: Classic and contemporary approaches to modeling biochemical reactions. Genes Dev 2010;24:1861–1875.

21 Bogdal MN, et al: Levels of pro-apoptotic regulator Bad and anti-apoptotic regulator Bcl-xL determine the type of the apoptotic logic gate. BMC Syst Biol 2013;7:67.

22 Luciani F, et al: A mathematical model for the immunosenescence. Riv Biol 2001;94:305–318.

23 Yang T, et al: Mathematical modeling of left ventricular dimensional changes in mice during aging. BMC Syst Biol 2012;6(suppl 3):S10.

24 Dyson J, Villella-Bressan R, Webb G: A spatial model of tumor growth with cell age, cell size, and mutation of cell phenotypes. Math Model Nat Phenom 2007;2:69–100.

25 Figge MT, et al: Deceleration of fusion-fission cycles improves mitochondrial quality control during aging. PLoS Comput Biol 2012;8:e1002576.

26 Kowald A, Kirkwood TB: Mitochondrial mutations and aging: random drift is insufficient to explain the accumulation of mitochondrial deletion mutants in short-lived animals. Aging Cell 2013;12:728–731.

27 Proctor CJ, Gray DA: GSK3 and p53 – is there a link in Alzheimer's disease? Mol Neurodegener 2010;5:7.

28 Tang MY, et al: Experimental and computational analysis of polyglutamine-mediated cytotoxicity. PLoS Comput Biol 2010;6:e1000944.

29 Proctor CJ, Kirkwood TB: Modelling cellular senescence as a result of telomere state. Aging Cell 2003;2:151–157.

30 Portugal RD, Land MG, Svaiter BF: A computational model for telomere-dependent cell-replicative aging. Biosystems 2008;91:262–267.

31 Needham CJ, et al: A primer on learning in Bayesian networks for computational biology. PLoS Comput Biol 2007;3:e129.

32 Chen R, et al: Dynamic Bayesian network modeling for longitudinal brain morphometry. Neuroimage 2012;59:2330–2338.

33 Hardy S, Robillard PN: Modeling and simulation of molecular biology systems using petri nets: modeling goals of various approaches. J Bioinform Comput Biol 2004;2:595–613.

34 Tomar N, et al: An integrated pathway system modeling of *Saccharomyces cerevisiae* HOG pathway: a Petri net based approach. Mol Biol Rep 2013;40:1103–1125.

35 Schlatter R, et al: ON/OFF and beyond – a Boolean model of apoptosis. PLoS Comput Biol 2009;5:e1000595.

36 Machado D, et al: Modeling formalisms in systems biology. AMB Express 2011;1:45.

37 Pogson M, et al: Introducing spatial information into predictive NF-kappaB modelling – an agent-based approach. PLoS One 2008;3:e2367.

38 Mc Auley MT, et al: Nutrition research and the impact of computational systems biology. J Comput Sci Syst Biol 2013;6:271–285.

39 Le Novère N, Hucka M, Mi H, et al: The systems biology graphical notation. Nat Biotechnol 2009;27:735–741.

40 Hoops S, et al: COPASI – a COmplex PAthway SImulator. Bioinformatics 2006;22:3067–3074.

41 Mc Auley MT, et al: A mathematical model of aging-related and cortisol induced hippocampal dysfunction. BMC Neurosci 2009;10:26.

Mark T. Mc Auley
Faculty of Science and Engineering
Thornton Science Park, University of Chester,
Chester, CH2 4NU (UK)
E-Mail m.mcauley@chester.ac.uk

Yashin AI, Jazwinski SM (eds): Aging and Health – A Systems Biology Perspective.
Interdiscipl Top Gerontol. Basel, Karger, 2015, vol 40, pp 49–62 (DOI: 10.1159/000364929)

How Does the Body Know How Old It Is?
Introducing the Epigenetic Clock Hypothesis

Joshua Mitteldorf

Department of EAPS, Massachusetts Institute of Technology, Cambridge, Mass., USA

Abstract

Animals and plants have biological clocks that help to regulate circadian cycles, seasonal rhythms, growth, development and sexual maturity. If aging is not a stochastic process of attrition but is centrally orchestrated, it is reasonable to suspect that the timing of senescence is also influenced by one or more biological clocks. Evolutionary reasoning first articulated by G. Williams suggests that multiple, redundant clocks might influence organismal aging. Some aging clocks that have been proposed include the suprachiasmatic nucleus, the hypothalamus, involution of the thymus, and cellular senescence. Cellular senescence, mediated by telomere attrition, is in a class by itself, having recently been validated as a primary regulator of aging. Gene expression is known to change in characteristic ways with age, and in particular DNA methylation changes in age-related ways. Herein, I propose a new candidate for an aging clock, based on epigenetics and the state of chromosome methylation, particularly in stem cells. If validated, this mechanism would present a challenging but not impossible target for medical intervention. © 2015 S. Karger AG, Basel

To many readers, it will seem like a strange idea that aging proceeds under control of a biological clock. If you think in terms of the body accumulating damage over time, then there is no master clock, no separate record of the state of the body's age – there are only the various parts of the body in various states of disrepair at any given time.

A recurring theme from research in animal models over the past two decades is that aging is not a passive accumulation of cellular damage, but is actively regulated at the level of the whole organism. However, even this regulation need not imply the existence of a biological clock; it may be merely that the activity of the body's repair mechanisms is modulated by an internal calculation based on external cues, so that the damage might accumulate at a variable rate without the body having to follow a scheduled program.

This chapter is adapted from an article of the same title published in *Biochemistry* (Mosc) [2013;78:1048–1053].

One motivation for thinking about a clock is a kind of Pascal's Wager[1] for the gerontologist: If there is an aging clock, then it suggests a convenient target for medical intervention that will have a highly leveraged effect on aging and disease. Speculation about the existence of an aging clock is interesting because, if such a thing does exist, it might be possible not just to slow the rate of aging, but to act directly on the clock, to set it back.

But Pascal's Wager (concerning the existence of God) appears to modern sensibilities to be a form of wishful thinking and an inversion of causal logic. We need a better reason for exploring the premise of an aging clock than the fact that it would be a boon for gerontologists if it turns out to be true.

Reasons to Believe in an Aging Clock

Here are four arguments in favor of an aging clock, which we shall explore in some detail presently. (1) In some animal models, interventions are known that do not simply slow the pace of decline, but actually cause reversion to a younger, more robust state. (2) Gene expression is known to change with age, including characteristic profiles that seem to be associated with senescence. (3) Attrition of telomeres seems in some ways to act like an aging clock. (4) Evidence that aging is an adaptive evolutionary program implies the existence of an aging clock.

(1) Many experimental interventions are known, for a given species, that cause the individual to revert to a younger state. Carrion beetles can be starved until they regress to a larval stage. Renewal of feeding causes them to mature again, and starvation can induce a repetition of the cycle through multiple 'lifetimes' [1]. The coelenterate *Turritopsis* has been observed to perform a similar feat outside the lab, in a natural setting [2, 3]. Lab mice have been rejuvenated with telomerase [4, 5] and with blood factors [6, 7]. In experiments with flies that are switched from a fully fed diet to caloric restriction (CR) in mid-life [8], it is found that the flies' mortality curve jumps quickly from the fully fed curve to the CR curve, as if they had been on CR from an early age. This suggests that, at least concerning those traits involved in mortality, CR does not merely slow the pace of future decline, but induces a change to a younger metabolic state.

Fahy [9] describes several more intriguing examples. Some of these might be conceived simply as upregulation of repair mechanisms, such that damage is temporarily being repaired faster than it accumulates. But even this conservative interpretation suggests that the body repairs itself more efficiently at younger ages, and this implies that the body 'knows how old it is' and chooses an age-appropriate efficiency of repair.

(2) Gene expression is now routinely profiled with DNA microarrays. Differences between late-life and early-life gene expression have been catalogued in several differ-

[1] Blaise Pascal (1523–1562) argued that we ought to believe in God for the following reason: If we believe in God and it turns out that our belief is erroneous, the consequence is trivial, but if we fail to believe in God and it turns out that God really exists, then the consequence is eternal damnation.

ent species [10–12]. In theory, these differences might be accounted for as the body's response to different levels of damage, rather than an age-dependent program; however, the nature of the changes with age suggests that they may be a cause of aging rather than a response to aging [12, 13]. For example, inflammation seems to be upregulated and immune function suppressed (in mice) by genes expressed late in life [12].

Early expositions of the pleiotropic theories for evolution of aging [14] were formulated before anything was known about the cell's and the body's elaborate controls over gene expression. In Williams's early conceptions of the fundamental pleiotropic mechanism, he imagined that if a gene was beneficial early in life, the body was stuck with it, even if its operation became detrimental to fitness later in life. We now know that this naïve version of pleiotropy is inconsistent with fundamentals of biochemistry. In particular, there is far more DNA devoted to transcription controls than there is genetic material that codes directly for proteins; and timing of gene expression is a fundamental element of all biological systems. All of development and maturation is controlled by biological clocks, so it is completely plausible that biological clocks are *available* for control of aging if we believe it possible that natural selection could have led in that direction.

(3) Short telomeres are associated with higher mortality and shorter life expectancy in many species, and the correlations persist when age is statistically factored out. Since telomeres shorten throughout the life span, and telomerase expression seems to be modulated in a manner consistent with regulation of life span, the idea of telomeres as an aging clock has been attractive to a number of authors [15–18]. A full section of the present chapter will be devoted to this hypothesis.

(4) I have collected evidence elsewhere [19, 20] that aging is a group-selected Darwinian adaptation, selected for its own sake. Others who have advocated this position explicitly include Skulachev [21], Bredesen [22], Goldsmith [23], Longo [24], Clark [25], Travis [26], Martins [27], Libertini [28] and Bowles [29]. Kenyon and Barja have indicated to me privately that sections of their submitted articles concerning programmed aging have been deleted by peer reviewers as a condition of publication. Many more scientists routinely adopt an implicit assumption that aging is programmed. But most evolutionists find this proposition implausible because the individual fitness cost of aging is high and the group-selected benefit is not confined to genetic kin. A brief summary of the reasons to believe that aging is an explicit evolutionary adaptation follows, and the reader is referred to my chapter in *The Future of Aging* [20] for details:

- Many of the genes that regulate aging (TOR, IGF, DAF/FOXO) have been conserved since the dawn of eukaryotic life [30]. All other such highly conserved genes have been protected by natural selection because they form an essential core to life processes. Natural selection has evidently treated aging as a core life process.
- Hormesis: The fact that life span can be readily extended under genetic control when conditions are most harsh and challenging (e.g., starvation) indicates that the body is 'holding back' on life span at times when the environment is more favorable [31, 32].

- Breeding animals for longevity does not necessarily impair fertility, as is demanded by popular theories based on pleiotropy or trade-offs [33]. In fact, many single-gene mutations are known to extend life (especially in worms), for which no major cost has yet been identified [34].
- One-celled eukaryotes are subject to two modes of programmed death: apoptosis [35] and replicative senescence [25, 36, 37]. This fact in itself vitiates the classical theoretical contention that it is impossible for programmed death to evolve, requiring as it would an implausible triumph of group selection over individual selection. Both apoptosis and replicative senescence have been conserved, so that they continue to play a role in the aging of higher organisms, including humans [38, 39].
- Many semelparous plants and animals exhibit manifestly programmed death [40–42], providing further counter-examples to the claim that affirmative selection for death is excluded on theoretical grounds.

Once we accept the premise that aging is programmed, it follows that the body must actively track its age (in a manner flexibly responsive to environmental cues) in order to initiate senescence at a characteristic age. The process must be governed by one or more master clocks.

Modulation of the Aging Clock Suggests a Demographic Purpose

The aging clock does not measure strict time, but is flexible in response to environmental conditions. The way in which the aging clock responds to the environment is highly suggestive of an adaptive purpose that helps us to understand the evolution of aging generally.

It is not possible to make sense of this picture if we imagine that life histories have been shaped only by natural selection for individual fitness. For the individual, living longer is always better. Theory predicts that all life histories should emulate the lobster or the sequoia tree, growing larger and more fertile and more robust against major causes of mortality with each passing year.

Instead, in iteroparous animals, we see a fixed life span, and the length of life varying *inversely* with hardship and environmental challenge, especially hunger. When the death rate from starvation is high, the death rate from aging is low, and vice versa. This is the well-known CR effect, but the same is also true of other environmental challenges: physical duress, infections, temperature extremes, and toxins in small amounts all lead paradoxically to longer life spans. The phenomenon is called *hormesis* [31, 32] and its reality has been established over the last two decades, after facing substantial skepticism of the initial accounts [43].

The demographic impact of senescence is thus to level the death rate in good times and bad, by imposing a higher mortality burden when the body is least stressed and lower just when it would appear to be metabolically most difficult to preserve the soma and avoid aging – under conditions of physical stress and starvation. By damping the most extreme variations in death rate, senescence makes possible the persistence and stability of ecosystems. Without aging, we might imagine a Tragedy of the Commons [44], with predator species competing viciously for their prey, and predator/prey population cycling much deeper than is actually observed in nature [45, 46].

The variability of the aging clock, and the existence of the clock itself, might be understood in terms of natural selection at the level of ecosystems. Ecosystems built on species that have a programmed life span are less likely to suffer overshoot, instability and collapse than if the species have indeterminate life span, limited only by starvation, predation, and epidemic infections. I have argued elsewhere [45, 46] that ecological homeostasis is a major target of natural selection, tempering and counterbalancing the pressure toward higher individual reproductive fitness.

Aging Clocks

If a program for aging were designed by a human engineer, it would be based on a central (flexible) time-keeping mechanism; but this is not necessarily the system that natural selection has bequeathed us. In particular, there is long-term group selection in favor of aging, but strong short-term individual selection against it. In order to shield affirmative aging mechanisms from dismantlement at the hand of individual selection, it is likely that evolution has embedded them below the surface, and deployed redundant time-keeping [14, 46]. A single master clock would be easily hijacked by individual selection. We might expect to find several interdependent and redundant clocks for aging.

Under the paradigm of programmed aging, senescence may be a continuation of development, and we might suspect that whatever controls the timing of development has been extended to regulate the timing of aging. But neither a developmental clock nor an aging clock has yet been discovered. The closest thing we know of is cellular senescence, based on telomere length [47]. But there are some organisms that age, in which telomerase is freely expressed, telomeres remain long, and thus the telomere clock is not operating. Examples include some rodents [48], bats [49] and pigs [50]. These considerations make it more likely that there is another clock and that it is somehow 'hidden in the works' of metabolism, so that it would not have been obvious to investigators thus far.

My hypothesis, proposed below, is that gene expression itself forms a kind of aging clock. Time is maintained within the signal networks of metabolism, and a running record of the organism's age is imprinted in the methylation state of the genome, as well as transcription factors and other regulators of gene expression. Gene expression products are part of a signal cascade that affects all aspects of metabolism, but that also feeds back (e.g. through methyl transferases) to increment the clock.

I will first briefly survey known biological clocks, including aging clocks, and discuss prospects for medical interventions that might manipulate them. These include thymic involution, the suprachiasmatic nucleus, the hypothalamus, and replicative senescence. The latter, based on telomere attrition, has been the subject of intense research in the past decade, with promising developments ongoing. Finally, I argue based on evidence and logic that the methylation state of the genome, within stem cells in particular, may be a promising place to look for a stored record of organismic age that informs the body's growth, development and senescence.

Cellular Senescence and the Telomere Clock

Telomeres in stem cells (and hence in their somatic progeny) suffer attrition over a human lifetime because they lose a few hundred base pairs with each cell division. The primary means by which telomere length might be restored is through the enzyme *telomerase*, which includes an enzyme which crawls along a chromosome end, and also an RNA template for copying the repetitive sequence. Telomerase is expressed copiously during early stages of an embryo's development, but very little telomerase is expressed after the individual passes beyond embryonic development. Hence, telomeres are permitted to shorten with age, even though the gene for telomerase remains (unexpressed) in the nucleus of every cell in the body.

The hypothesis that cellular senescence represents a primary aging clock was promoted by West, culminating in a popular book published in 2003 [17]. That same year saw Cawthon's actuarial study [39], associating leukocyte telomere length with mortality in humans. In the years since then, evidence has accumulated for the importance of telomere length in aging of birds and mammals including humans, and several herbal extracts that are claimed to address telomeric aging have reached the market. Meanwhile, several companies are researching more potent telomerase activators in the belief that this is a path to substantial extension of the human life span.

The Cawthon results forced many researchers to consider for the first time the possibility that people could be dying for lack of telomerase. Association between telomere length and life expectancy was confirmed in 3 studies of animals in the wild [51–53]. The question remained open whether longer telomeres were a marker or a cause of life expectancy. This question has been addressed with animal studies. Telomeres have been extended by adding ectopic copies of the telomerase gene, by genetically programming the expression of telomerase via a tamoxifen switch, and by oral administration of a plant-derived compound that promotes telomerase expression. Life span extension has been detected in worms, mice and rats.

Joeng et al. [54] created a strain of *Caenorhabditis elegans* worms with longer telomeres using not telomerase, but a telomere-binding protein called HRP-1. Life span was extended by 19% by this intervention. The result was unexpected because telomeres do not erode over the life span of *C. elegans*. In fact, the adult worms are postmitotic: there are no stem cells, no replenishment of tissues during a single worm's lifetime. It should not be possible for telomeres to function as an aging clock. Life extension of the HRP-1 worms was dependent on the presence of DAF-16, an upstream modifier of aging that is thought to be a master regulator of dauer formation in response to environmental hardships.

Tomás-Loba [55] first demonstrated life extension in mice using a strain that was engineered with extra copies of the telomerase (TERT) gene. Because it was widely believed that telomerase expression could cause cancer, they used mice that were cancer resistant via modified *p53*. These mice lived 40% longer than controls, and markers of senescence such as inflammation, glucose tolerance and neurological measures appeared on a delayed schedule. This result was unexpected because wild-type mice express telomerase copiously, and their telomeres are long enough to last through several lifetimes without obvious effects on health and longevity [56–58].

There have also been some negative indications, casting doubt on the hypothesis that telomeres are an aging clock. Telomere length is widely variable in newborns [59], with a standard deviation about 7% of the mean. Telomere length predicts mortality and life expectancy less well with advancing age, with correlation disappearing for ages >80 [60, 61]. A few studies [62, 63] have reported a tumorigenic effect of telomerase, suggesting that the evolutionary purpose for rationing telomerase is the opposite of an aging clock.

There are two known mechanisms by which telomeric aging might lead to senescence of the organism and greater risk of mortality. First, as stem cells suffer from telomere attrition, they slow down in replacing the skin and blood and immune cells that are constantly turning over. Second, cells with short telomeres enter a senescent phase where they emit toxic signals, including proinflammatory cytokines [64, 65]. Hence, it is not surprising to see cellular senescence emerge as a primary driver of senescent phenotypes.

The role of telomere length as a cellular aging clock recapitulates a similar function in some *protoctista*. In paramecia, for example, telomerase is not expressed during mitosis, but only during conjugation. Hence, paramecia may reproduce clonally through a few hundred generations before their telomeres become shortened and they enter a senescent state, losing viability. They are compelled to conjugate, blending their genomes sexually with a partner cell, or they cease to be able to reproduce. Hence, the rationing of telomerase serves to enforce an imperative to share genes, and helps to assure that the local population remains diverse, and thus robust. Telomeric senescence serves to damp a winner-take-all form of individual selection, and enhanced diversity is insurance against excessive specialization, which might offer temporary advantage [25].

Other Known Biological Clocks

Circadian biological clocks have been widely studied and are partially understood. There is evidence for an annual clock that contributes, along with environmental cues, to patterns of migration and hibernation; but mechanisms have not been identified. And the timing of growth, reproductive maturity, and senescence remain more mysterious yet.

The mammalian circadian rhythm is known to be regulated from the suprachiasmatic nucleus, a small structure in the middle of the brain [66]. A chemical mechanism (based on peroxiredoxins) has recently been discovered that might underlie all circadian clocks [67]. The cycle is responsive to light and dark, and the intrinsic period is close enough to 24 h that circadian timing becomes entrained with diurnal cycles.

It has been proposed that seasonal cycles in mammals are mediated through a response to light in the pineal gland. The mechanism is thought to be independent of the circadian clock [68]. Again, there is an intrinsic cycle time that is modulated by temperature and duration of daylight.

In female mammals, onset of puberty is controlled by a single chemical signal: gonadotropin-releasing hormone (GnRH). But the timing of this trigger is controlled in turn by a complex calculation, based in neural as well as hormonal mechanisms. 'The anatomical development of the GnRH secretory system occurs relatively early in life, and the synthetic capacity is present well before puberty in that GnRH mRNA expression reaches adult levels' [69]. Timing responds to olfactory cues, stress, fat reserve, activity, season of the year, and other stimuli. The workings of this clock remain mysterious.

Aging responds to these same cues, and perhaps others. There is reason to believe that the aging clock mechanism is at least as complex as the developmental clock. Though aging is programmed, it may not be programmed in a simple way. This accounts for the challenge that aging has posed for research and medical intervention. The fact that aging progresses under genetic control suggests a promising approach to anti-aging interventions, and yet the complexity of the timing mechanism has slowed the pace of progress.

Although relationships between the circadian clock and the aging clock have been documented, these are not such as to suggest that the aging clock depends directly on a count of circadian cycles. For example, dysregulation of the circadian clock *in either direction* leads to accelerated aging in flies [70, 71].

The Neuroendocrine Theory of Aging was proposed by Vladimir Dilman in 1954 [72]. Homeostatic control of hormone secretions is supported by the hypothalamus, and different hormonal levels are maintained as is appropriate for different stages of growth and development. Dilman's hypothesis was that the trajectory of changes in the hypothalamus has a kind of momentum ('hyperadaptosis') that carries forward after maturity and results in 'dysregulation' that characterizes the aging phenotype. The Neuroendocrine Theory is an early precedent for the Epigenetic Theory described below. 'The life span, as one of the cyclic body functions regulated by "biological clocks", would undergo a continuum of sequential stages driven by nervous and endocrine signals' [73]. But Dilman did not frame this theory within the context of an adaptive program shaped by natural selection, and therefore the concept of a biological clock sits uncomfortably within its narrative.

The Immunologic Theory was proposed by Roy Walford in 1962 [74, 75]. The proliferation of immune cells in the blood constitutes a kind of clock, which becomes dysfunctional as the number of cells multiply. The body's cells are mutating as the number of different immune memory cells is multiplying. Chance coincidences result in the immune system attacking self with increasing frequency over time. Walford noted how many diseases of old age have a relationship to autoimmunity, but never connected this to programmed aging. He saw thymic involution as an independent cause of immune failure, and perhaps another aging clock.

Epigenetic Clock Hypothesis

In the fall of 2012, an article [13] appeared by Adiv Johnson and a diverse team of scientists from the US and Europe pulling together evidence that the methylation state of the genome is related to the body's age, and proposing methylation as an appropriate target for anti-aging research. I would extend their proposal to argue that, if we believe there is an aging clock, the methylation state of the genome (especially in stem cells, because of their persistence) is logically the first place to look for its 'clock dial'. Seeking a system of global signals that affects the metabolic state of the entire body, we would look as far upstream as possible. Upstream takes us to gene expression. Further up, there may be signals that affect gene expression globally, but these, too, are products of genes, and hence they can be regarded as part of a self-modifying program for gene expression. If there is not in evidence another separate clock which feeds down to affect gene expression, then it is logical to assume that this self-modifying program functions as a clock in its own right.

We know that gene expression changes with age, and that this has the potential to affect all aspects of the metabolism and the aging phenotype. If there is an aging clock, then its output must be transduced so as to affect gene transcription. Merging the clock into the transcription state of the genome would be the most economical implementation of a clock mechanism, obviating the need for a separate record of the age state of the body. Gene transcription is affected by transient signaling, and also by more persistent epigenetic markers. The most important of these persistent markers is the genome's methylation pattern. The 'methylome' contains information that is both programmable and persistent. Cytosine (the 'C' in ACGT) is one of the four nucleic acid residues that form chromosomal DNA. Within the DNA molecule, cytosine can accept a methyl group to form 5-methylcytosine, and this suppresses transcription locally where methylation has occurred in gene promoters [76]. Methylation patterns tend to be copied along with DNA replication, and they can even last through several generations as a form of epigenetic inheritance [77].

An epigenetic clock has the potential to regulate growth, development and sexual maturity, as well as aging. If no other clock has been discovered that controls the timing of both development and aging, then our default hypothesis ought to be that the epigenetic state of the genome is its own clock.

The methylation state of the genome is also self-modifying in the sense that transcription of methyl transferases and related enzymes creates the mechanism for feeding back upon the methylation state. This feedback implies the basis for a clock mechanism. Genes that are transcribed today create the metabolic environment that cascades into signals that reconfigure the methylation state and program the genes that will be transcribed tomorrow.

The above constitutes a general, theoretical argument for epigenetic state as an aging clock. There are also specific experimental results that point in this direction:

- Gene expression profiles change substantially with age. There is reason to believe that an individual with youthful gene expression is functionally a youthful individual [13].
- Methylation has about the right degree of persistence. We know that methylation contains epigenetic information that is passed on in a soft way when DNA is replicated.
- In general, methylation decreases with age (though there are characteristic regions that become hypermethylated). Hypomethylation has been associated with 'frailty' and markers of biological age [78].
- Fruit flies with an extra copy of the methyl transferase *dnmt2* in their genome lived 58% longer than control flies. Conversely, flies engineered to be +/− for *dnmt2* had lifespans 25% shorter [79].
- Similar experiments with mice yield more nuanced results. Early, unreproduced studies reported that methylation was actually higher in *dnmt1* +/− mice than in +/+ controls [80, 81]. More recently, neural deficiencies and low bone densities, increasing with age, have been reported associated with engineered *dnmt1* deficiencies [82].
- In monozygotic twins, methylation patterns are similar when young, but diverge over time [83]. This suggests a stochastic component that may account for diverse dysregulations associated with aging.

Age is determined almost certainly by the detailed pattern of methylation and other epigenetic markers, not simply the crude quantity of methylation. And yet there is evidence that senescing cells are characterized by progressive demethylation, so that chromosomes in younger cells tend to be more methylated than older cells [84, 85]. The possibility that demethylation may be an aging clock was first proposed by Bowles [29], based on the fact that 'aging is accompanied by DNA demethylation [86–88]. In fact, the animal genome loses practically all 5-methylcytosines during the life, the rate of the loss being inversely proportional to maximal lifespan of the species [88]. The same occurs in cell cultures, again the rate being inversely proportional to the cell lifespan (Hayflick limit) [88–90].'

Skulachev also notes a connection between oxidation, which has often been recognized as a stochastic marker for aging, and methylation. In his schema, oxidation is a more fundamental aging clock (rather than demethylation leading to oxidation, as I argue here). '[O]xidation by ROS of the guanine DNA residues to 8-hydroxyguanine strongly inhibits methylation of adjacent cytosines [91]. Antioxidants, on the other hand, cause DNA hypermethylation [92]. According to Panning and Jaenisch [93], DNA hypomethylation activates Xist gene expression in X chromosome, which correlates with a dramatic stimulation of apoptosis. All these observations may be summarized by the following chain of age-related events: ROS → DNA de-methylation → apoptosis → aging [90].'

Drugs targeted to sirtuins [94] have found some early success in extending lifespan by indiscriminate silencing of gene expression. Sirtuins' mechanism is mediated via histone deacetylation rather than methylation, but the ease with which simple silencing of genes could extend lifespan is suggestive. Also, protein-restricted and methionine-restricted diets retard aging [95], presumably by dialing down expression of many genes indiscriminately. Methionine is the 'start codon', essential to the initiation of all gene transcription.

Testing the Hypothesis: Medical Implications

This hypothesis – that the body's age is stored within the cell nucleus as a methylation pattern – suggests a program of research, and an anti-aging strategy. Interventions based on methylation will require both a detailed automated reading of the methylation state of the genome, and a means of transcribing a youthful profile into chromosomes in vitro.

The former is already fairly well developed. Heyn et al. [85] report transcription of the methylome using microarrays. The latter may be far more challenging. Methyl transferases are able to methylate targeted genes, but details of the biochemistry that guides the transferases to their target is not yet understood [13].

More accessible might be interventions to increase methylation broadly. This is a life extension strategy that has been made to work in flies [79] but not yet in mammals. S-adenosyl methionine (SAMe) is the basic methyl donor of all eukaryotes. Simple supplementation with SAMe has been found to relieve arthritis and depression symptoms [96], and SAMe has been shown to protect methylation levels in radiation-challenged mice [97], but SAMe has not been found to extend lifespan in rodents. Johnson et al. [13] catalogue some nutrients that have been associated with reduced methylation, but none with enhanced methylation. They stress that we are yet at an early stage of knowledge concerning the relationship between methylation and aging, and it is not yet proven even that alterations of the methylation state are a cause and not simply a product of aging. Nevertheless, they propose methylation as a promising avenue for foundational and clinical research, and I concur. Since this article was written, CRISPR technology (Clustered, Regularly Interspaced, Short Palindromic Repeats) has advanced rapidly. Just a few years ago, the technique was developed to target a place in the genome for gene editing. It is now expected that CRISPR will also provide a handle for dictating epigenetics [98], turning genes on [99] and off [100] at will. If this comes to pass, CRISPR may provide a dramatic shortcut, cutting through the complex biochemistry of gene expression and permitting us to target genes for promotion or repression. There are already several known genes with presumptive anti-aging benefits (e.g. GDF11, Klotho, possibly oxytocin, melatonin), and others that are overexpressed late in life with detrimental consequences (e.g. NFκB, TGF-β, possibly LH and FSH).

References

1 Beck SD, Bharadwaj RK: Reversed development and cellular aging in an insect. Science 1972;178:1210–1211.

2 Piraino S, et al: Reversing the life cycle: Medusae transforming into polyps and cell transdifferentiation in *Turritopsis nutricula*. Biol Bull 1996;90:302–312.

3 Barinaga M: Mortality: overturning received wisdom. Science 1992;258:398–399.

4 Jaskelioff M, et al: Telomerase reactivation reverses tissue degeneration in aged telomerase-deficient mice. Nature 2011;469:102–106.

5 Bernardes de Jesus B, et al: Telomerase gene therapy in adult and old mice delays aging and increases longevity without increasing cancer. EMBO Mol Med 2012;4:691–704.

6 Conboy IM, et al: Rejuvenation of aged progenitor cells by exposure to a young systemic environment. Nature 2005;433:760–764.

7 Katcher H: Studies that shed new light on aging. Biochemistry (Mosc) 2013;78:1061–1070.

8 Mair W, et al: Demography of dietary restriction and death in *Drosophila*. Science 2003;301:1731–1733.

9 Fahy G: Precedents for the biological control of aging: postponement, prevention and reversal of aging processes; in Fahy GM, et al (eds): Approaches to the Control of Aging: Building a Pathway to Human Life Extension. New York, Springer, 2010.

10 Jin W, et al: The contributions of sex, genotype and age to transcriptional variance in *Drosophila melanogaster*. Nat Genet 2001;29:389–395.

11 Golden TR, Melov S: Microarray analysis of gene expression with age in individual nematodes. Aging Cell 2004;3:111–124.

12 Sharman EH, et al: Effects of melatonin and age on gene expression in mouse CNS using microarray analysis. Neurochem Int 2007;50:336–344.

13 Johnson AA, et al: The role of DNA methylation in aging, rejuvenation, and age-related disease. Rejuvenation Res 2012;15:483–494.

14 Williams G: Pleiotropy, natural selection, and the evolution of senescence. Evolution 1957;11:398–411.

15 Harley CB, Villeponteau B: Telomeres and telomerase in aging and cancer. Curr Opin Genet Dev 1995;5:249–255.

16 Fossel M: Reversing Human Aging. New York, Harpercollins, 1997.

17 West MD: The Immortal Cell. New York, Doubleday, 2003, p 244.

18 Aviv A, Bogden JD: Telomeres and the arithmetic of human longevity; in Fahy GM, et al: The Future of Aging: Pathways to Human Life Extension. New York, Springer, 2010, pp 573–586.

19 Mitteldorf J: Aging selected for its own sake. Evol Ecol Res 2004;6:1–17.

20 Mitteldorf J: Evolutionary origins of aging; in Fahy GM, et al (eds): The Future of Aging: Pathways to Human Life Extension. New York, Springer, 2010.

21 Skulachev VP: Programmed death phenomena: from organelle to organism. Ann N Y Acad Sci 2002;959:214–237.

22 Bredesen DE: The non-existent aging program: how does it work? Aging Cell 2004;3:255–259.

23 Goldsmith T: The Evolution of Aging. Crownsville, Azinet, 2003, 2008.

24 Longo VD, Mitteldorf J, Skulachev VP: Programmed and altruistic ageing. Nat Rev Genet 2005;6:866–872.

25 Clark WR: Reflections on an unsolved problem of biology: the evolution of senescence and death. Adv Gerontol 2004;14:7–20.

26 Travis JM: The evolution of programmed death in a spatially structured population. J Gerontol A Biol Sci Med Sci 2004;59:301–305.

27 Martins AC: Change and aging senescence as an adaptation. PLoS One 2011;6:e24328.

28 Libertini G: An adaptive theory of the increasing mortality with increasing chronological age in populations in the wild. J Theor Biol 1988;132:145–162.

29 Bowles JT: The evolution of aging: a new approach to an old problem of biology. Med Hypotheses 1998;51:179–221.

30 Guarente L, Kenyon C: Genetic pathways that regulate ageing in model organisms. Nature 2000;408:255–262.

31 Forbes V: Is hormesis an evolutionary expectation? Funct Ecol 2000;14:12–24.

32 Masoro EJ: The role of hormesis in life extension by dietary restriction. Interdisc Top Gerontol 2007;35:1–17.

33 Leroi A, Chippindale AK, Rose MR: Long-term evolution of a genetic life-history trade-off in *Drosophila*: the role of genotype-by-environment interaction. Evolution 1994;48:1244–1257.

34 Arantes-Oliveira N, Berman JR, Kenyon C: Healthy animals with extreme longevity. Science 2003;302:611.

35 Fabrizio P, et al: Superoxide is a mediator of an altruistic aging program in *Saccharomyces cerevisiae*. J Cell Biol 2004;166:1055–1067.

36 Clark WR: Sex and the Origins of Death. Oxford, Oxford University Press, 1998, p 208.

37 Clark WR: A Means to an End: The Biological Basis of Aging and Death. Oxford, Oxford University Press, 1999, p 234.

38 Behl C: Apoptosis and Alzheimer's disease. J Neur Trans 2000;107:1325–1344.

39 Cawthon RM, et al: Association between telomere length in blood and mortality in people aged 60 years or older. Lancet 2003;361:393–395.

40 Barry TP, et al: Free and total cortisol levels in semelparous and iteroparous chinook salmon. J Fish Biol 2001;59:1673–1676.

41 Kirkwood TB, Thomas BL, Melov S: On the programmed/non-programmed nature of ageing within the life history. Curr Biol 2011;21:R701–R707.

42 Goldsmith TC: Arguments against non-programmed aging theories. Biochemistry (Mosc) 2013;78:971–978.

43 Calabrese EJ: Toxicological awakenings: the rebirth of hormesis as a central pillar of toxicology. Toxicol Appl Pharmacol 2005;204:1–8.

44 Hardin G: The tragedy of the commons. Science 1968;162:1243–1248.

45 Mitteldorf J: Chaotic population dynamics and the evolution of aging: proposing a demographic theory of senescence. Evol Ecol Res 2006;8:561–574.

46 Mitteldorf J: Adaptive aging in the context of evolutionary theory. Biochemistry (Mosc) 2012;77:716–725.

47 Mitteldorf J: Telomere biology: cancer firewall or aging clock? Biochemistry (Mosc) 2013;78:1054–1060.

48 Seluanov A, et al: Telomerase activity coevolves with body mass not lifespan. Aging Cell 2007;6:45–52.

49 Wang L, McAllan BM, He G: Telomerase activity in the bats *Hipposideros armiger* and *Rousettus leschenaultia*. Biochemistry (Mosc) 2011;76:1017–1021.

50 Fradiani P, et al: Telomeres and telomerase activity in pig tissues. Biochimie 2004;86:7–12.

51 Pauliny A, et al: Age-independent telomere length predicts fitness in two bird species. Mol Ecol 2006;15:1681–1687.

52 Haussmann MF, Winkler DW, Vleck CM: Longer telomeres associated with higher survival in birds. Biol Lett 2005;1:212–214.

53 Bize P, et al: Telomere dynamics rather than age predict life expectancy in the wild. Proc Biol Sci 2009;276:1679–1683.

54 Joeng KS, et al: Long lifespan in worms with long telomeric DNA. Nat Genet 2004;36:607–611.

55 Tomás-Loba A, et al: Telomerase reverse transcriptase delays aging in cancer-resistant mice. Cell 2008;135:609–622.

56 Wynford-Thomas D, Kipling D: Telomerase: cancer and the knockout mouse. Nature 1997;389:551–552.

57 Chang S: Modeling aging and cancer in the telomerase knockout mouse. Mutat Res 2005;576:39–53.

58 Mendelsohn AR, Larrick JW: Ectopic expression of telomerase safely increases health span and life span. Rejuvenation Res 2012;15:435–438.

59 Okuda K, et al: Telomere length in the newborn. Pediatr Res 2002;52:377–381.

60 Bischoff C, et al: No association between telomere length and survival among the elderly and oldest old. Epidemiology 2006;17:190–194.

61 Kimura M, et al: Telomere length and mortality: a study of leukocytes in elderly Danish twins. Am J Epidemiol 2008;167:799–806.

62 Stewart SA, et al: Telomerase contributes to tumorigenesis by a telomere length-independent mechanism. Proc Natl Acad Sci 2002;99:12606–12611.

63 Bagheri S, et al: Genes and pathways downstream of telomerase in melanoma metastasis. Proc Natl Acad Sci 2006;103:11306–11311.

64 Campisi J: Senescent cells, tumor suppression, and organismal aging: good citizens, bad neighbors. Cell 2005;120:513–522.

65 Rodier F, Campisi J: Four faces of cellular senescence. J Cell Biol 2011;192:547–556.

66 Klein DC, Moore RY, Reppert SM: Suprachiasmatic Nucleus: The Mind's Clock. Oxford, Oxford University Press, 1991.

67 Edgar RS, et al: Peroxiredoxins are conserved markers of circadian rhythms. Nature 2012;485:459–464.

68 Danks H: How similar are daily and seasonal biological clocks? J Insect Physiol 2005;51:609–619.

69 Ebling FJ: The neuroendocrine timing of puberty. Reproduction 2005;129:675–683.

70 Kumar S, Mohan A, Sharma VK: Circadian dysfunction reduces lifespan in *Drosophila melanogaster*. Chronobiol Int 2005;22:641–653.

71 Dubrovsky YV, Samsa WE, Kondratov RV: Deficiency of circadian protein CLOCK reduces lifespan and increases age-related cataract development in mice. Aging (Albany) 2010;2:936.

72 Dilman VM, Dean W: The neuroendocrine theory of aging and degenerative disease. Pensacola, Center for Bio Gerontology, 1992.

73 Weinert BT, Timiras PS: Invited review: theories of aging. J Appl Physiol 2003;95:1706–1716.

74 Walford RL: The Immunologic Theory of Aging. Gerontologist 1964;4:195–197.

75 Walford RL: The Immunologic Theory of Aging. Gerontologist 1964;4:195–197.

76 Cooney C, Lawren B: Methyl magic: Maximum Health through Methylation. Kansas, Andrews McNeel Publishing, 1999.

77 Jablonka E, Raz G: Transgenerational epigenetic inheritance: prevalence, mechanisms, and implications for the study of heredity and evolution. Q Rev Biol 2009;84:131–176.

78 Bellizzi D, et al: Global DNA methylation in old subjects is correlated with frailty. Age 2012;34:169–179.

79 Lin M-J, et al: DNA methyltransferase gene dDnmt2 and longevity of *Drosophila*. J Biol Chem 2005;280:861–864.

80 Yung R, et al: Unexpected effects of a heterozygous Dnmt1 null mutation on age-dependent DNA hypomethylation and autoimmunity. J Gerontol A Biol Sci Med Sci 2001;56:B268–B276.

81 Ray D, et al: Aging in heterozygous Dnmt1-deficient mice: effects on survival, the DNA methylation genes, and the development of amyloidosis. J Gerontol A Biol Sci Med Sci 2006;61:115–124.

82 Liu L, et al: Insufficient DNA methylation affects healthy aging and promotes age-related health problems. Clin Epigenet 2011;2:349–360.

83 Fraga MF, et al: Epigenetic differences arise during the lifetime of monozygotic twins. Proc Natl Acad Sci USA 2005;102:10604–10609.

84 Wilson VL, Jones PA: DNA methylation decreases in aging but not in immortal cells. Science (New York) 1983;220:1055.

85 Heyn H, et al: Distinct DNA methylomes of newborns and centenarians. Proc Natl Acad Sci USA 2012;109:10522–10527.

86 Vanyushin B, et al: The 5-methylcytosine in DNA of rats. Gerontology 1973;19:138–152.

87 Wilson VL, et al: Genomic 5-methyldeoxycytidine decreases with age. J Biol Chem 1987;262:9948–9951.

88 Mazin A: Genome loses all 5-methylcytosine a life span. How is this connected with accumulation of mutations during aging? (in Russian). Mol Biol (Mosk) 1993;27:160.

89 Mazin A: Loss of total 5-methylcytosine from the genome during cell culture aging coincides with the Hayflick limit (in Russian). Mol Biol (Mosk) 1993; 27:895.

90 Skulachev VP: Aging and the programmed death phenomena; in Nyström T, Osiewacz HD (eds): Model Systems in Aging. Berlin, Springer, 2004, pp 191–238.

91 Weitzman SA, et al: Free radical adducts induce alterations in DNA cytosine methylation. Proc Natl Acad Sci USA 1994;91:1261–1264.

92 Romanenko EB, Alessenko AV, Vanyushin BF: Effect of sphingomyelin and antioxidants on the in vitro and in vivo DNA methylation. Biochem Mol Biol Int 1995;35:87.

93 Panning B, Jaenisch R: DNA hypomethylation can activate Xist expression and silence X-linked genes. Genes Dev 1996;10:1991–2002.

94 Kelly G: A review of the sirtuin system, its clinical implications, and the potential role of dietary activators like resveratrol: part 1. Altern Med Rev 2010; 15:245–263.

95 Zimmerman JA, et al: Nutritional control of aging. Exp Gerontol 2003;38:47.

96 Baldessarini RJ: Neuropharmacology of S-adenosyl-L-methionine. Am J Med 1987;83(suppl 1):95–103.

97 Batra V, Sridhar S, Devasagayam TPA: Enhanced one-carbon flux towards DNA methylation: effect of dietary methyl supplements against γ-radiation-induced epigenetic modifications. Chem Biol Interact 2010;183:425–433.

98 Friedland AE, et al: Heritable genome editing in *C. elegans* via a CRISPR-Cas9 system. Nat Methods 2013;10:741–743.

99 Ranganathan V, et al: Expansion of the CRISPR-Cas9 genome targeting space through the use of H1 promoter-expressed guide RNAs. Nat Commun 2014;5:4516.

100 Kiani S, et al: CRISPR transcriptional repression devices and layered circuits in mammalian cells. Nat Methods 2014;11:723–726.

Joshua Mitteldorf
Department of EAPS
Massachusetts Institute of Technology
Cambridge, MA 02138 (USA)
E-Mail josh@mathforum.org

Mitteldorf

Yashin AI, Jazwinski SM (eds): Aging and Health – A Systems Biology Perspective.
Interdiscipl Top Gerontol. Basel, Karger, 2015, vol 40, pp 63–73 (DOI: 10.1159/000364930)

The Great Evolutionary Divide: Two Genomic Systems Biologies of Aging

Michael R. Rose · Larry G. Cabral · Mark A. Philips · Grant A. Rutledge · Kevin H. Phung · Laurence D. Mueller · Lee F. Greer

Department of Ecology and Evolutionary Biology, University of California, Irvine, Calif., USA

Abstract

There is not one systems biology of aging, but two. Though aging can evolve in either sexual or asexual species when there is asymmetric reproduction, the evolutionary genetics of aging in species with frequent sexual recombination are quite different from those arising when sex is rare or absent. When recombination is rare, selection is expected to act chiefly on rare large-effect mutations, which purge genetic variation due to genome-wide hitchhiking. In such species, the systems biology of aging can focus on the effects of large-effect mutants, transgenics, and combinations of such genetic manipulations. By contrast, sexually outbreeding species maintain abundant genetic polymorphism within populations. In such species, the systems biology of aging can examine the genome-wide effects of selection and genetic drift on the numerous polymorphic loci that respond to laboratory selection for different patterns of aging. An important question of medical relevance is to what extent insights derived from the systems biology of aging in model species can be applied to human aging.   © 2015 S. Karger AG, Basel

With the advent of whole-genome DNA sequencing and other genomic technologies, biology is now revealing much about the genetic foundations of life that were once obscure. Here, we discuss the relevance of some of these early genomic findings for the systems biology of aging, as well as the design and interpretation of experimental research on aging.

We structure our discussion of these issues in terms of two distinct kinds of population genetic systems: largely asexual reproduction or sexual reproduction with limited outcrossing (system 1), and sexual reproduction with consistent outbreeding (system 2). These two systems have very different consequences for genomes, particularly the relationship between genetic variation and organismal function, including aging. Their differing patterns of genetic variation and adaptation produce striking differences in the relationship between genetic variation and aging, with profound consequences for the theoretical explanation and experimental analysis of the genomic foundations for the systems biology of aging.

Rarely Sexual Evolutionary Genetics: Theoretical Expectations for Aging Genomics

Complete asexuality is evolutionarily exceptional. The prokaryotes, for example, may lack frequent mendelian recombination, but nonetheless have parasexual mechanisms of recombination: conjugation, transduction, and transformation. Such parasexual genetic exchange can be largely or wholly removed from laboratory evolution paradigms that use well-understood prokaryotes such as *Escherichia coli* [1]. But in nature, parasexuality doubtless occurs with enough frequency to produce novel genotypes at a greater frequency than mutation within clonal lineages can achieve. Early work showed that bacterial genetic diversity is higher than previously thought, suggesting infectious transfer of genes brings in new genetic variants which are not necessarily purged by directional selection [2]. Indeed, when there is frequency-dependent selection as well as transposable, phage, and plasmid elements, even bacterial populations can maintain significant genetic variation [3].

However, if parasexuality introduces new genetic variation into a population at a sufficiently low rate, then it is roughly like mutation with respect to its impact on functional genetics. That is, if mendelian segregation and recombination among variant alleles do not usually occur in each generation, then conventional theory for clonal selection can address the evolutionary effects of the favorable genetic variants by genetic exchange which is sufficiently rare. Note that this does *not* mean that the genomics of non-mendelian populations would not be significantly different with parasexual genetic exchange. Compared to wholly asexual populations, there should be more genetic variation within parasexual or infrequently outcrossing sexual populations. But if we assume that beneficial genetic variation is only introduced rarely by parasexual recombination or amphimixis, then the low frequency at which genetic variation is introduced over time gives rise to an evolutionary process of adaptation mathematically comparable to that of adaptation sustained by rare favorable mutations subject to clonal selection.

Thus, it is reasonable to examine the scenarios for adaptation and for aging that arise with this evolutionary genetic 'system 1' in terms of the formal theory and experimental findings already available for asexual systems with intermittent favorable mutations. The theory of such evolutionary systems is relatively powerful [4]. Small-effect beneficial variants will often be lost because of genetic drift effects near the fixation boundary, as they will in mendelian populations. Large-effect beneficial variants are much more likely to escape from such sampling effects and sweep to fixation, purging genetic variation along the way due to a genome-wide, rather than local, hitchhiking effect [cf. 5]. Thus, the process of adaptation itself will tend to purge genetic variation from such systems, genome-wide. The chief exceptions to this genomic purging arise in cases where selection is frequency-dependent, such as arises with micro-niche partitioning among genotypes [e.g. 6]. It is unclear how often such frequency-dependent balancing selection leads to the maintenance of genetic variation in asexual or parasexual populations in nature [but see 3].

The effects of this evolutionary genetic system 1 on the genomics of aging can be interpreted in terms of standard hamiltonian theory for age-specific selection [7, 8]. If there is strictly symmetrical division of the organism during reproduction, then aging is not expected to evolve. It is important to note that the common asexual reproductive system of fission does not ensure such strict symmetry. In both bacteria and eukaryotes, cytologically asymmetrical fission is well known to allow the evolution of aging [9]. Even when cytological asymmetry is not obvious, bacteria like *E. coli* nevertheless have asymmetrical partitioning of waste products after fission, which can lead to the evolution of aging [10]. Nonetheless, under good conditions strictly fissile single-cell species like *Schizosaccharomyces pombe* are known to be free of aging [11], as are Hydra kept under good conditions [12], as expected by hamiltonian theory [7–9].

But when there is sufficient asymmetry between products of asexual reproduction, the problem of declining age-specific forces of natural selection is expected to always lead to the evolution of aging in hamiltonian evolutionary theory [7–9]. Dissent from this view has been offered [13], but numerical simulations of asexual evolution suggest that such alternative theory for the evolution of aging in age-structured populations is incorrect [8]. Only the formalism of Hamilton has been shown to be intimately tied with fitness in an age-structured population under some well-defined conditions [8, 14]. No such connection has been made for alternative non-hamiltonian analyses [13].

At the genomic level, mutations with solely deleterious effects at later ages might drift to fixation, a process commonly called mutation accumulation [8, 9, 14]. But because of the intermittent purging of genetic variation by selective sweeps, such drift effects are particularly unlikely for asexual systems. A more likely genetic mechanism for the evolution of aging in asexual species is antagonistic pleiotropy [8, 9, 14, 15], in which mutations that have early life benefits sweep to fixation despite deleterious later-life effects, because the latter will have little effect on fitness compared to the former.

There are some significant issues that arise with evolutionary genomic system 1 for aging which are too often neglected: genotype-by-environment interaction ('GxE'), rate of approach to evolutionary equilibrium, and age-independent beneficial substitutions. Furthermore, these issues potentially interact with each other. Both adaptation and aging are environment specific, thanks to GxE. That is, a mutant that is beneficial in one environment can be deleterious in another environment. This is well understood for adaptation in evolutionary research, but less appreciated for aging, even though aging is nothing more or less than age-dependent loss of adaptation.

When mutants have beneficial effects that are age-independent, they foster the evolution of indefinitely sustained survival and reproduction, producing the phenomenon of post-aging plateaus after the cessation of aging [8]. But such evolutionary effects are dependent on the number of generations that a population has spent in a particular environment. The approach to mutation-selection equilibrium in asexual populations decelerates over evolutionary time, with some evidence of continuing adaptation as late in evolution as 50,000 generations [16], despite assiduous maintenance of a stable culture regime. This means that asexual populations with age structure will be subject to a protracted process of adaptation with later ages particularly subject to a lack of beneficial effects, which produces aging.

Rarely Sexual Evolutionary Genetics: Implications for the Systems Biology of Aging

The forgoing general theoretical and experimental points have great salience for the design and interpretation of gerontological systems biology research using rarely outcrossing species like *Saccharomyces cerevisiae* and *Caenorhabditis elegans*. In these species, longevity mutants can be isolated by random mutational screens or by screening extant mutant stocks isolated for other purposes. It is not difficult to find mutants with greatly increased longevities in rarely outcrossing species [e.g. 17]. There are several evolutionary genetic mechanisms by which such increased longevity can arise.

First, since these longevity mutants are not generally isolated from asexual stocks that have long adapted to laboratory conditions, unlike those of Lenski [16], they may be mutants that enhance survival and reproduction generally in the evolutionarily novel environment supplied in a particular laboratory. In effect, then, these mutants may be generally beneficial in the specific lab environments in which they are screened. That is, they are broadly adaptive. As such, they are not specifically 'aging mutants', even if they increase longevity. There is nothing wrong with the careful study of such genetic mechanisms of adaptation, so long as the experimenters doing so understand that they are studying the genomic foundations of adaptation in general, not something that is notably specific to aging. Evolutionary biologists welcome new recruits to the study of the evolutionary genetics of adaptation, particularly when those new recruits are aware of the strong evolutionary theory that pertains to adaptation.

Second, other longevity mutants might involve cases of striking antagonistic pleiotropy, in which early reproduction or competitive ability are sacrificed in the mutant in exchange for a striking gain in adult longevity, at least in the particular laboratory environment employed by the experimenter. Van Voorhies et al. [18] have argued that this is the case for a number of *C. elegans* longevity mutants. However, the ability to obtain such antagonistic pleiotropy mutants does not demonstrate that aging evolved in such species specifically because of the 'wild-type' sequences that have been altered in these longevity mutants.

Rose·Cabral·Philips·Rutledge·Phung·Mueller·Greer

Evolutionary Genetics of Outcrossing Mendelian Populations: General Genomic Findings

Over the last 30 years, research on the evolutionary genetics of mendelian populations that usually outbreed has been dominated by a 'neoclassical' consensus. Classical evolutionary genetic theory presumed there was very little segregating mendelian variation [19], with rare beneficial mutations that sweep through the genomes of such populations providing the foundation of adaptation. With the discovery of massive genetic variation in outbred mendelian populations beginning in the late 1960s, classical theory was replaced by a neoclassical variant with the following assumptions: (a) a great deal of strictly neutral genetic variation evolves by genetic drift; (b) deleterious genetic variation arises by mutation but is kept at a low frequency by purifying selection, and (c) rare beneficial mutations arise and sweep to fixation [20]. Like the classical theory of evolutionary genetics, neoclassical theories imply that there will be very little segregating genetic variation genome-wide that can readily respond evolutionarily to novel forms of selection. Instead, rare beneficial mutations of large effect are expected to dominate the process of adaptation in outbred mendelian populations, as they are supposed to in the evolution of strictly asexual populations [1].

Long opposed to both classical and neoclassical theories, balance and neobalance theories for evolutionary genetics have hypothesized that there is abundant genetic variation affecting components of fitness. The difference between balance and neobalance theories is analogous to the difference between classical and neoclassical theories. Like neoclassical theory, present-day neobalance theory adds to traditional balance theory the possibility that there is a significant amount of neutral or weakly deleterious genetic variation across the genome. The key difference is that neobalance theory supposes that balancing mechanisms of selection, such as frequency-dependent selection and antagonistic pleiotropy, frequently allow the maintenance of functional genetic variants at high frequencies. Such abundant functional genetic variation is then expected to allow immediate responses to selection at many sites across the genome.

Recent experimental studies of the genome-wide response to selection in laboratory mendelian populations suggest that the neobalance theory has much more validity than has been supposed heretofore. For example, Burke et al. [21] found numerous sites in the genome that respond strikingly to selection for earlier reproduction and faster aging in *Drosophila melanogaster*. Our recent unpublished research has found even more numerous sites in the *Drosophila* genome that respond to selection on aging, thanks to superior replication to that of earlier studies of the genomics of the response to selection in mendelian populations. Thus, there can be a very large number of sites in the genome of a sexual species that respond to selection, implicating the action of widespread balancing selection which sustains functional genetic diversity in outbred populations of species like those of *Drosophila*.

Evolutionary Genetics of Outcrossing Mendelian Populations: Implications for Their Systems Biology of Aging

From a hamiltonian standpoint, aging is the decline in age-specific adaptation due to the declining forces of natural selection, as illustrated by, among other things, the cessation of aging after these forces stop declining [8]. But the genomic machinery that underlies this decline in age-specific adaptation is evidently very different between largely asexual and sexual species, if the neobalance view of the evolutionary genetics of outbred sexual species is correct. If the neoclassical view is correct, there is *no* such differentiation between the genomics of aging in asexual and sexual species: aging in both would depend primarily on selective sweeps of alleles with antagonistic pleiotropic effects that force aging as a price of improved adaptation at earlier ages. The neoclassical theory chiefly has different implications for outbred mendelian populations compared to rarely sexual species with respect to mutation accumulation, because the purging of genetic variation that occurs with selective sweeps is expected to be localized across the genome in outbred species, rather than global [5, 20]. Our view, however, is that this neoclassical view of the evolution of aging in sexual populations is unlikely to be correct, given the observed genomics of the evolution of aging in *Drosophila* [21].

There is thus a clear divide in the genomic underpinnings of aging between largely asexual species and outbreeding mendelian species, a division of great significance for the systems biology of aging. In particular, the kinds of mutants that are generated by mutagenesis and then identified in screens for increased life span do not generally correspond to those identified by us when resequencing outbred laboratory-evolved *D. melanogaster* populations with increased longevity.

This type of finding is apparently not confined to aging research. The prosaic but much studied character of *Drosophila* bristle number likewise shows little correspondence between loci identified by mutagenesis and those identified from studies of genetic variation in wild populations [e.g. 22]. Thus, the functional genomics of outbred system 2 populations are not expected to correspond to the genetics of large-effect mutants on theoretical grounds, and they do not appear to correspond so far in studies of *Drosophila* where a direct comparison can be made. Naturally, this conclusion can only be offered for *Drosophila* at this point, but the history of genetics suggests that many *Drosophila* findings are likely to generalize.

The Two Kinds of Evolutionary Genomics and the Systems Biology of Aging

Systems Biology of Aging in Largely Asexual or Inbred Populations

Considered in, of, and for themselves, evolutionary genetic system 1 species like *E. coli* or *S. cerevisiae* can be usefully studied using large-effect mutants, as shown in figure 1.

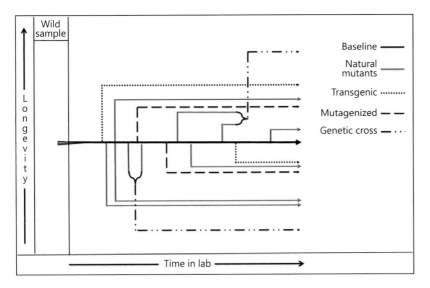

Fig. 1. In largely asexual or inbred species, the systems biology of aging is appropriately analyzed using large-effect mutants, transgenics, etc. This research focus arises naturally from the way aging evolves in such species. In the figure, genetic strains of little heterozygosity are indicated by thin lines, with the different types of genetic manipulation indicated by different line fonts.

It is exactly such large-effect mutations which will be important determinants of the evolution of aging in such species, because when they are sufficiently beneficial for fitness they will sweep rapidly toward fixation, purging genetic variation across the rest of the genome. Once they have swept through a population from a largely asexual species, such substitution will change the activity of at least some, and potentially many, gene products from elsewhere in the genome. Such effects on gene activity should be identifiable from gene expression assays, from microarray characterization of translated proteins or from genome-wide transcriptomic assays. Furthermore, such substitutions could then pave the way for further, now beneficial, epistatic substitutions elsewhere in the genome. And such epistatic combinations of substitutions could likewise be studied usefully, whether by direct genetic manipulation or experimental evolution [23].

From data concerning both mutants and the alleles fixed by selective sweeps, including gene expression and other downstream phenotypic effects, it should be possible to build a systems biological model for how some large-effect DNA sequence changes work their way through complex molecular networks to produce specific patterns of adaptation and aging in inbreeding or asexual organisms. In particular, the specific details of the antagonistic pleiotropy likely to be involved in natural selection against most longevity mutants would be a natural theme of such research, as well as any genotype-by-environment dependence in the appearance and disappearance of such effects [see 18]. Overall, this seems like an eminently feasible research project.

Significant challenges face the systems biology of aging in outcrossing mendelian species. Many of these challenges have already arisen in the study of chronic diseases in human GWAS research. It has proven very difficult to identify the majority of specific SNP changes that are responsible for the heritability of such chronic disorders as type 2 diabetes, Crohn's disease, or Alzheimer's disease, although the impact of ApoE and ACE variants on longevity stands out as a notable exception [24]. But the total amount of genetic variation for longevity that is explained by such loci is very small.

As we have already argued in more general terms [25], experimental evolutionary genomics offers a powerful way to begin developing a systems biology for aging in outbreeding mendelian species. Since there are many sites in the genome with sequence variation that affect aging, such variation can be teased out using experimental evolution with resequencing. Experimental evolution can readily change the rate of aging, and even the age of onset of late-life aging plateaus [8–9]. With enough replication of selection lines and enough generations of selection, it is possible to probe the entire genomes of organisms like *D. melanogaster* for the genomic sites that affect aging [cf. 21]. Furthermore, it is then straightforward to use orthology to identify corresponding loci in other outbred mendelian species that affect aging or chronic disease in those species. Figure 2 provides a schematic for this research strategy.

In this sense, the systems biology of aging is easier to probe genome-wide in outbred mendelian populations than it is in inbred or asexual species. It requires much less experimental effort to rapidly screen entire genomes to find a large number of genomic regions that are involved in the response to selection for different patterns of aging. However, recently Tenaillon et al. [23] have shown that very large-scale experimental evolution in asexual populations can come up with comparable genome-wide screening. In their case, they applied high-temperature selection to more than 100 distinct populations for 2,000 generations. By resequencing samples from this huge collection of populations after selection, they were able to identify genomic regions that were repeatedly subject to sequence change as a result of their specific selection protocol. But sustaining longevity or later-life fertility selection among some hundred or more clonal lines over thousands of generations seems like a prohibitively difficult task.

In any case, we already know that this strategy of genomic-site identification works in studies of the experimental evolution of aging in outbred mendelian populations with much less replication [21]. The question is then what to do with information about the many genomic sites that evidently play a role in the system 2 genetics of aging, with respect to the systems biology of aging.

An obvious next step is the use of genome-wide gene expression assays of populations with different patterns of aging over a range of different ages and different environments in order to infer the regulatory interconnections across the numerous sites in the genome. For further resolution, tissue-specific gene expression could also be used, particularly in conjunction with the kind of functional physiological differ-

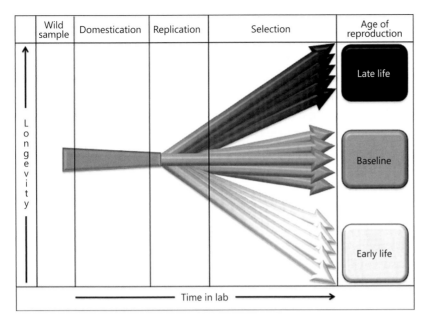

Fig. 2. Experimental evolution readily shifts patterns of aging when Hamilton's forces of natural selection have their declining phases shifted by changing the time at which outbred sexual laboratory populations are cultured [8]. Such populations are differentiated at many sites across the genome [21]. The systems biological analysis of aging in such settings needs to proceed along very different lines from that which is appropriate for asexual or inbred species.

entiation associated with the evolution of aging in *Drosophila*. Then there is the possibility of metabolomic analysis of whole organisms and specific tissues or organs. Such downstream genomic and functional data would naturally be extremely complex, requiring advanced machine learning software to parse. Any viable systems biological model for aging obtained by this experimental methodology would naturally be extremely complex in turn, a model that would be opaque to human intuition at the detailed level. Nonetheless, it is possible to parse even very complex numerical models using well-developed bioinformatics tools like sensitivity analysis, in which model parameters are systematically varied in simulations for their quantitative impact on model predictions.

This in turn raises the question of how to test such a systems biological model. In our laboratory, we have a number of populations of *D. melanogaster* that have evolved a wide range of differences in DNA sequence, gene expression, and downstream phenotypes. If we were to assemble a systems biological model for aging based on genome-wide changes in sequence and gene expression among a subset of our populations, then we could test the validity of the analysis of such a model by determining whether corresponding variation in populations that were *not* used to construct the initial systems biology model nonetheless shows the predicted patterns of parametric sensitivity.

Application to the Systems Biology of Aging in Humans

Many scientists who work on aging are chiefly interested in patterns of aging in the particular taxonomic group(s) that they study, but most people are more interested in the application of such research to the biomedical problem of aging, as illustrated by the articulate declamations of figures like Aubrey de Grey.

Thus, the question of applying systems biological insights into the genomics of aging among model organisms to the human case is a natural concern. Surprisingly, a large fraction of the loci identified as relevant to aging in *D. melanogaster* have orthologs in the human genome, and some of those loci in turn have been associated with human chronic disease, particularly when human GWAS databases are probed in light of genomic results [26]. Likewise, the various longevity loci identified in studies of largely asexual species can be tested for their relevance to human genomic databases for aging-associated diseases and disorders.

In the end, the question of the medical merit of the systems biology of aging in model species for human aging is empirically answerable with orthology. We are already confident that genomic data from similarly outbreeding mendelian species are, at a minimum, relevant to *understanding* the systems biology of aging in our species, but it has not yet been determined how useful such understanding will be for developing interventions that might *ameliorate* human aging.

Acknowledgements

We thank Jennifer E. Briner for her careful reading and editing of the manuscript.

References

1 Lenski RE, Rose MR, Simpson SC, Tadler SC: Long-term experimental evolution in *Escherichia coli*. I. Adaptation and divergence during 2,000 generations. Am Nat 1991;138:1315–1341.

2 Levin BR: Periodic selection, infectious gene exchange and the genetic structure of *E. coli* populations. Genetics 1981;99:1–23.

3 Levin BR: Frequency-dependent selection in bacterial populations. Philos Trans R Soc Lond B Biol Sci 1988;319:459–472.

4 Nagylaki T: Introduction to Theoretical Population Genetics. Berlin, Springer, 1992.

5 Maynard Smith J, Haigh J: The hitch-hiking effect of a favourable gene. Genet Res 1974;23:23–35.

6 Rainey PB, Travisano M: Adaptive radiation in a heterogeneous environment. Nature 1998;394:69–72.

7 Hamilton WD: The moulding of senescence by natural selection. J Theor Biol 1966;12:12–45.

8 Mueller LD, Rauser CL, Rose MR: Does Aging Stop? New York, Oxford University Press, 2011.

9 Rose MR: Evolutionary Biology of Aging. New York, Oxford University Press, 1991.

10 Lindner AB, Madden R, Demarez A, Stewart EJ, Taddei F: Asymmetric segregation of protein aggregates is associated with cellular aging and rejuvenation. PNAS 2008;105:3076–3081.

11 Coelho M, Dereli A, Haese A, Kühn S, Malinovska L, Desantis ME, Shorter J, Alberti S, Gross T, Tolić-Nørrelykke IM: Fission yeast does not age under favorable conditions, but does so after stress. Curr Biol 2013;23:1844–1852.

12 Martinez DE: Mortality patterns suggest lack of senescence in hydra. Exp Gerontol 1998;33:217–225.

13 Baudisch A: Hamilton's indicators of the force of selection. PNAS 2005;102:8263–8268.

Rose · Cabral · Philips · Rutledge · Phung · Mueller · Greer

14 Charlesworth B: Evolution in Age-Structured Populations. New York, Cambridge University Press, 1980.

15 Rose MR: Life-history evolution with antagonistic pleiotropy and overlapping generations. Theor Pop Biol 1985;28:342–358.

16 Lenski RE: Evolution in action: a 50,000-generation salute to Charles Darwin. Microbe 2011;6:30–33.

17 Klass MR: A method for the isolation of longevity mutants in the nematode Caenorhabditis elegans and initial results. Mech Ageing Dev 1983;22:279–286.

18 Van Voorhies WA, Curtsinger JW, Rose MR: Do longevity mutants always show trade-offs? Exp Gerontol 2006;41:1055–1058.

19 Lewontin RC: The Genetic Basis of Evolutionary Change. New York, Columbia University Press, 1974.

20 Burke MK: How does adaptation sweep through the genome? Insights from long-term selection experiments. Proc R Soc 2012;B279:5029–5038.

21 Burke MK, Dunham JP, Shahrestani P, Thornton KR, Rose MR, Long AD: Genome-wide analysis of a long-term evolution experiment with Drosophila. Nature 2010;467:587.

22 Gruber JD, Genissel A, Macdonald SJ, Long AD: How repeatable are associations between polymorphisms in achaete-scute and bristle number variation in Drosophila? Genetics 2007;175:1987–1997.

23 Tenaillon O, Rodríguez-Verdugo A, Gaut RL, McDonald P, Bennett AF, Long AD, Gaut BS: The molecular diversity of adaptive convergence. Science 2012;335:457–461.

24 Schachter F, Fauredelanef L, Guenot F, Rouger H, Froguel P, Lesueurginot L, Cohen D: Genetic associations with human longevity at the ApoE and ACE loci. Nat Genet 1994;6:29–32.

25 Rose MR, Mueller LD, Burke MK: New experiments for an undivided genetics. Genetics 2011;188:1–10.

26 Rose MR, Long AD, Mueller LD, Rizza CL, Matsagas KC, Greer LF, Villeponteau B: Evolutionary nutrigenomics; in Fahy GM, West M, Coles LS, Harris SB (eds): The Future of Aging: Pathways to Human Life Extension. New York, Springer, 2010.

Michael R. Rose
Department of Ecology and Evolutionary Biology, University of California
321 Steinhaus Hall
Irvine, CA 92697-2525 (USA)
E-Mail mrrose@uci.edu

Yashin AI, Jazwinski SM (eds): Aging and Health – A Systems Biology Perspective.
Interdiscipl Top Gerontol. Basel, Karger, 2015, vol 40, pp 74–84 (DOI: 10.1159/000364932)

Development and Aging: Two Opposite but Complementary Phenomena

Bruno César Feltes · Joice de Faria Poloni · Diego Bonatto

Laboratory of Molecular and Computational Biology, Department of Molecular Biology and Biotechnology,
Center of Biotechnology, Federal University of Rio Grande do Sul, Rio Grande do Sul, Brazil

Abstract

Aging is a consequence of an organism's evolution, where specific traits that lead to the organism's development eventually promote aged phenotypes or could lead to age-related diseases. In this sense, one theory that broadly explored development and its association to aging is the developmental aging theory (DevAge), which also encompasses most known age-associated theories. Thus, we employed different systems biology tools to prospect developmental and aging-associated networks for human and murine models for evolutionary comparison. The gathered data suggest a model where proteins related to inflammation, development, epigenetic mechanisms and oxygen homeostasis coordinate the interplay between development and aging. Moreover, the mechanism also appears to be evolutionary conserved in both mammalian models, further corroborating the DevAge molecular model.   © 2015 S. Karger AG, Basel

A Brief View of the Two Sides of Life

Aging is accepted as a universal and irreversible process that gradually promotes physiological cellular decay and occurs in almost all living beings, from prokaryotes to eukaryotes. With aging as one of the main causes of disease development and tissue functionality impairment in mammals, different theories have been devised to describe and understand the causes and consequences of aging, including the generation of free radicals, the accumulation of DNA mutations, and the shortening of telomeres.

Nevertheless, understanding aging and its associated biological processes requires an answer to the question, 'why do we age?'. The antagonistic pleiotropic theory, which was first proposed by Williams in 1957 [1], provided the following first glimpse of a satisfactory explanation: aging is a consequence, a declining force of natural selection, where the same specific molecular mechanisms and traits that can benefit young organisms can also be deleterious during the course of life, ultimately leading to aged phenotypes. These traits can arise during early development, providing advantages in terms of organogenesis, morphogenesis and embryo survival, but they can also promote aging. This scenario can be exemplified by the oxidative stress induced by reactive oxygen species. Oxygen is necessary for proper tissue functionality, but long-term exposure is correlated with aging-associated diseases and cancer development that are attributed to the oxidation of proteins, lipids and nucleic acids [2]. Another example is insulin, which ensures proper fetal growth by regulating several metabolic processes, such as the formation of adipose tissue. When deregulated, however, the insulin-mediated signaling pathway also correlates with diabetes, which is commonly associated with aging and will be discussed further.

When and how an organism ages remains inconclusive, although it has been proposed that the aging process begins after an organism reaches its maximal reproductive capability. Nevertheless, several aging-associated pathologies are thought to result from genetic predisposition combined with specific environmental factors that become evident during the life span of an organism (e.g. Alzheimer's disease [3]). In addition, it is important to consider that environmental changes during pregnancy can induce differential developmental outcomes by altering the epigenome via changes in the methylation and acetylation of DNA and histones, irreversibly modifying gene expression patterns [4]. Thus, we need to advance our understanding of aging by analyzing other theories that encompass the above conditions, such as 'how are developmental outcomes connected to aging?'. In this sense, the following three theories/hypotheses fit into the criteria detailed above: the thrifty phenotype hypothesis (TPH) proposed by Hales and Barker [5, 6], the developmental origins of health and disease (DOHaD) theory proposed by Langley-Evans [4], and the developmental-aging (DevAge) theory proposed by Dilman in 1971 [7].

The first two hypotheses are complementary and are derived from the same principle. Aging results from genetic changes that occur during embryonic development and that are driven by environmental alterations, such as malnutrition of the mother or abnormalities in the placenta or in other maternal physiologies [4–6]. These alterations can significantly alter fetal growth and tissue development [4–6]. On the other hand, DevAge theory expands the scenarios encompassed by the TPH and DOHaD hypotheses and suggests that aging is part of the same molecular mechanism that promotes tissue development and embryonic maturation during development, continuing throughout adult life [3].

In summary, aging is a programmed molecular mechanism – a tradeoff *per se* – and is essential for the embryonic maturation that results in deleterious effects and aging phenotypes. Nonetheless, those theories, although supported by numerous trustworthy observations, still require more profound observations at the molecular level.

Common Mechanisms between Development and Aging Outcomes

Understanding complex molecular mechanisms, such as development and aging, and verifying their interconnectivity and how they are coordinated represent ideal challenges for computational modeling, where the use of networks can be applied for dynamic and global views of molecular interactions. Thus, our group developed a systems biology study to understand such mechanisms.

The Essential Understanding of Topological Parameters

To understand the relation between aging and development, we prospected two different protein-protein (PPI) networks, using the data available for *Homo sapiens* and *Mus musculus*. As expected, the networks (fig. 1) displayed nodes (proteins) related to development and aging (e.g. HOX, PAX, histone deacetylases, DNA methyltransferases, and sirtuins). The topology of each network was subsequently analyzed. We first calculated the following two major parameters: clustering and centralities (fig. 1).

Clustering is subject to the principle of a union between individual parts of a system that are directly or indirectly connected. This principle of organization is commonly observed in daily life, either in the way that we organize objects by function or resemblance or in our tendency to organize social life by affinity groups of beliefs or mentalities. In a biological network, clusters are characterized by high-density regions that normally act on a specific biological function or biochemical pathway [8].

Centralities are commonly used to analyze the most topologically relevant nodes in a given network [9]. The following two parameters were used: node degree and betweenness.

The first parameter (node degree) refers to nodes exhibiting node degree values above the average node degree value as hubs, which are defined as nodes with several

Fig. 1. A representation based on interatomic data illustrating how aged phenotypes are driven by programmed molecular mechanisms during development. Environmental conditions, such as the exposure to different pollutants and climate fluctuations contribute together with nutritional habits and/or substance abuse (e.g. tobacco smoke, and alcohol), to changes in maternal health and placental status, which affects different molecular mechanisms. The data gathered from the PPI networks illustrate the major altered mechanisms that include (but are not limited to) inflammation, glucose metabolism, epigenetic programming (e.g. CpG island methylation and histone modifications) and changes in O_2 status. The DevAge and DOHaD hypotheses are derived from the principle that the overstimulation or inhibition of those mechanisms can significantly alter fetal development and terminally promote age-associated diseases. Notably, proinflammatory cytokines are necessary for the initial signaling events that promote the formation of the embryo stem cell niche and the establishment of the maternal-fetal interface. However, these are the same mechanisms that are responsible for the development of various diseases. SC = Stem cell.

(For figure see next page.)

Feltes · de Faria Poloni · Bonatto

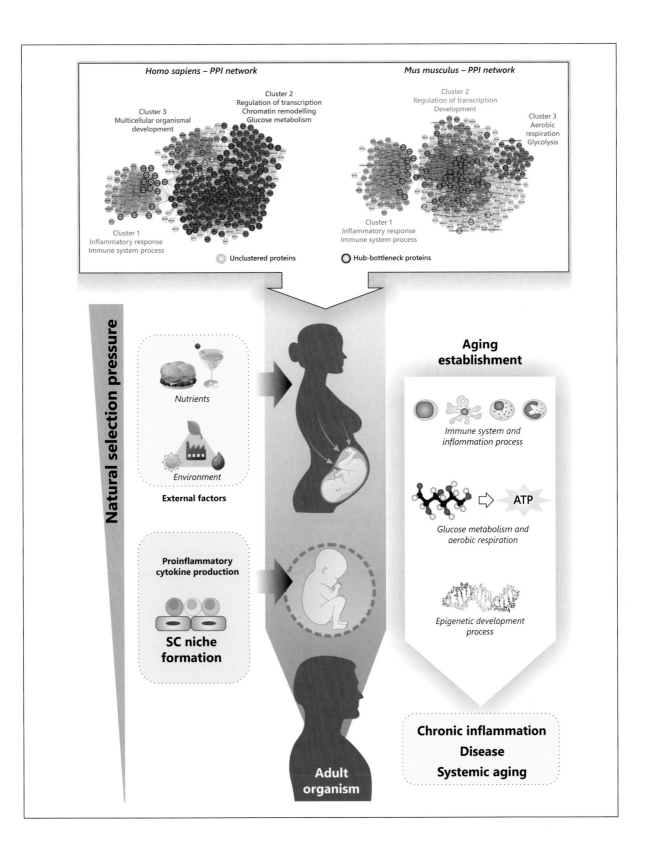

connections [9]. In contrast, betweenness is correlated with the flow of information that passes through a given node, including the number of connections (edges) that pass through a node, which are required for the connection of clusters and other nodes to the network [9]. Thus, proteins with a betweenness score above the network average betweenness value are named bottlenecks. The union of the two parameters results in so-called hub-bottleneck nodes that are crucial for network integrity [10] (fig. 1). These nodes were the main focus of our study in terms of centralities.

Finally, we prospect the associated biological processes present in our network.

Decoding the Networks: New Insights into DevAge and Its Relationship to Inflammation

As predicted, the networks (fig. 1) exhibited similar topologies, indicating their conservation among the two species, as well as the following similarly associated bioprocesses: inflammatory and immune system processes, and the regulation of development and glycolysis. Among the bioprocesses present, the *H. sapiens* network displayed chromatin remodeling, whereas the *M. musculus* network displayed aerobic respiration (fig. 1).

The fact that both of the networks displayed a close relationship with inflammation is interesting. Inflammation has been demonstrated to be intimately correlated with aging diseases such as Parkinson's, Alzheimer's, sarcopenia, osteoporosis and atherosclerosis [11]. The relationship between aging and inflammation is thought to represent one of the many causes of age-related phenotypes that characterize the so-called inflammaging theory (IT). The IT posits that the chronic inflammation that an organisms suffers over the course of its life span as a result of constant exposure to innumerous antigens results in reduced efficiency of the immune system (immunosenescence), which promotes the development of aging-associated and autoimmune diseases [11]. These diseases are predominantly caused by the overexpression of proinflammatory cytokines and inflammatory mediators [11]. Our centrality analysis (fig. 1) revealed that the hub-bottleneck subnetworks of both model networks contained several proinflammatory proteins, including interferon-γ (IFNG), tumor necrosis factor, and nuclear factor-κB. The *H. sapiens* network displayed even more inflammatory-related factors, such as interleukins (ILs) -2, -4, -6, -8, -10, -18 and -1β, which are crucial mediators of pro- and anti-inflammatory cascades, whereas the *M. musculus* network showed the presence of IL-2, -4, -6, -10, -18, -1a, -1b, -1r1, -10ra, -17a, -2ra, -2rb, -2rg [12]. Our networks (*H. sapiens* and *M. musculus*) also identified the caudal type homeobox 2 (CDX2), which is a predominant regulatory protein in intestinal development [12]. This protein contains a binding site for nuclear factor-κB, indicating a close relationship between the immune system, inflammatory response and development [12]. These observations reflect the significant conservation of the inflammation-associated nodes of both networks that were initially queried for development and aging.

Moreover, in adults, wound healing is a continual process that involves the following three distinct phases: inflammation, proliferation and tissue remodeling [13]. Active wounds are characterized by anoxic or hypoxic environments and anaerobic metabolism, which represent a possible stem cell niche [13]. Additionally, lymphatic circulation facilitates the infiltration of the interstitial space by inflammatory components and the production of inflammatory mediators such as IL-1, IL-6 and tumor necrosis factor-α, proteins which were present in our PPI network (fig. 1) [12, 13]. Finally, blood circulation is regenerated, reestablishing oxygen transport and nutrient supply, promoting tissue remodeling [13]. Thus, Aller et al. [13] argued that the tissue repair of the body of injuries during postnatal life might occur according to embryonic biochemical patterning in a manner similar to gastrulation, which recapitulates an ancestral biochemical mechanism. This process might also recapitulate embryonic ontogeny in injured tissues via a hypothetical trophic axis, which comprises the amniotic and yolk sacs [13].

The amniotic and yolk sacs are extraembryonic tissues that surround the embryo, which represent the amniotic and vitelline axes, respectively, and which contribute to the maintenance of the mesoderm that localizes within the interstitial space between them. Furthermore, the amniotic sac is responsible for the secretion of amnion-derived cellular growth factors and cytokines, which promotes mesenchymal and epithelial communication [13]. In addition, the amniotic fluid acts as an extracellular extension of the fetus by regulating changes in interstitial hydroelectrolytic potential [13]. Accordingly, the yolk sac is a membranous structure that is formed during gastrulation and contains hematopoietic progenitors called blood islands, which contain associated endothelial cells [13]. The differentiation of the yolk sac promotes hematopoiesis and the development of a primitive circulatory network [13]. Additionally, the yolk sac has the major function of embryo nutrition by accumulating amino acids, carbohydrates and lipids [13]. Thus, Aller et al. [13] argue that the repair of tissues in the body after injuries during postnatal life occurs according to embryonic biochemical patterns in a similar manner as gastrulation by utilizing an ancestral biochemical mechanism. This process could recapitulate embryonic ontogeny via a hypothetical trophic axis that consists of the amniotic and yolk sac in injured tissues [13].

Therefore, both the amniotic and yolk sac contribute to normal development and are essential for organogenesis. Their functions are integrated during intraembryonic mesoderm formation and are maintained as mesoderm-derived cells in the form of fibroblasts, which can be identified within postnatal connective tissues [13]. Thus, the inflammatory responses in injured tissues recapitulate an amniotic-vitelline phenotype; a hypoxic environment can provide the ideal conditions for the expansion and differentiation of stem cells, in addition to the recruitment of immune cells and the regeneration of circulation [13]. However, the overstimulation of inflammatory cytokines can impair embryo implantation, compromise proper maternal-fetus vascular interface, and prevent optimal nutritional inflow [12].

These observations are consistent with the TPH, DOHaD and DevAge theories, where such maternal alterations can alter embryonic development. Thus, the IT hypothesis could fall into DevAge theory context.

Decoding the Networks: DevAge and Its Relation to Epigenetics

Studies have demonstrated that during the 20th century, newborns with low birthweights also exhibited high blood pressure, a predisposition for coronary heart disease and the development of type 2 diabetes during adult life [14]. Such observations corroborate the TPH, which proposes that occasional changes in intrauterine conditions during embryonic development might induce fetal adaptations during nutrient deficiency by optimizing fetal energetic supplies [15]. However, organ development can be restricted during nutrient deprivation, impairing fetal growth [15]. Certain tissues (e.g. cardiac and neural tissues) are crucial for proper body functions, whereas others (e.g. osteogenic and muscular tissues) can display specific plasticities, exhibiting abnormal functions without affecting the short-term survival of the fetus [15].

In contrast to Barker's group, the predictive adaptive response (PAR) hypothesis proposed by Bateson and Gluckman [16] argues that a fetus exhibits the inherent ability to promote metabolic reorganization and to adapt to environmental stresses, such as nutrient deficiency [17]. Thus, environmental stresses during early life can provide a 'forecast' by dictating the environment conditions in which the organism will grow, promoting the development of an adequate phenotype for the stressors present in that environment [17]. Both the PAR and TPH hypotheses are associated with developmental plasticity in response to environment changes and with how this plasticity is correlated with adult health and disease. Developmental plasticity appears as an adaptive mechanism that is governed by a set of underlying molecular processes, such as epigenetic programming [18]. The association of these molecular mechanisms, how they are trigged during development, and how they affect adult life are explained by the DevAge and DOHaD theories.

Our PPI networks for both organisms (fig. 1) displayed several of the following proteins that are related to epigenetics: (a) the DNA methyltransferase (DNMT1), DNMT3A, DNMT3B; (b) class I histone deacetylase 1 (HDAC1), HDAC2 and HDAC3; (c) class III HDACs named sirtuin 1 (SIRT1) and SIRT2, and (d) p300, a histone acetyltransferase. All the proteins listed above are crucial for the maintenance of chromatin structure. Epigenetic studies have aimed to determine heritable changes in chromatin structure by analyzing histones and CpG island modifications, which induce conformational changes in chromatin, facilitating the binding of transcription machinery at specific promoter regions in different tissues. This phenomenon is commonly referred to as the 'on' (relaxed chromatin, available for the binding of transcription factors) or 'off' (condensed chromatin, unavailable for transcriptional machinery) state of a gene [18].

Moreover, it has been suggested that DNA methylation patterns (controlled by DNMTs during development) are modulated by the nutritional status of an organism, representing a potential regulator of phenotypes and tissue plasticity [18]. The DOHaD hypothesis supports all of these ideas and is consistent with the DevAge theory, which describes the interpolation of development and aging mechanisms and which assumes that fetal development is ruled by a set of key genes that are turned 'on' or 'off' by epigenetic mechanisms to induce phenotypic changes in response to the intrauterine environment and the maternal state. These changes might have pathophysiological relevance, contributing to the predisposition to disease or, in the less aggressive cases, to phenotypes that would only manifest during adulthood. Major epidemiological implications can include cardiovascular disease, type 2 diabetes, depression, osteoporosis and impaired cognitive functions [19]. Thus, developmental mechanisms are susceptible to natural selection of a regulatory set of genes that are essential for proper body and axis patterning and progenitor tissue specification [12, 20].

Decoding the Networks: Glucose Metabolism and Developmental Changes

In the PPI networks, we observed several proteins in both networks that are related to glucose and aerobic metabolism (in *M. musculus*), such as succinate dehydrogenase (SDHA and SDHB), aconitase (ACO1 and ACO2), and glycerol-3-phosphate dehydrogenase 1 and 2 (GPD1/GPD2). We also identified the insulin growth factor 1 in our analysis (IGF1, which was present in both model organisms).

Glucose metabolism and aerobic respiration are essential for appropriate fetal growth and normal adult metabolism [21]. Glucose utilization within the fetus is more dramatic than during late gestation, when the developing pancreas produces insulin upon glucose stimulation [21]. Aberrant fetal insulin secretion occurs in response to oscillating glucose concentrations. Interestingly, chronic hyper- and hypoglycemic conditions can alter glucose transporter concentrations [21], indicating that glucose transporters are altered to reduce glucose uptake during hyperglycemia, which can correlate with the development of insulin resistance [21]. In addition, altered insulin and glucose pathways can induce glucose uptake during hypoglycemic states, which is associated with intrauterine growth restriction (IUGR) [21]. Furthermore, IUGR is generally related to fetal hypoglycemia, which involves decreased insulin secretion and fetal pancreatic development [21]. These findings are related to B-cell dysfunction and reduced pancreatic endocrine tissue mass, which, over the long term, might increase the incidence of noninsulin-dependent diabetes mellitus [21]. In contrast, hyperglycemia is associated with protein glycation and increased production of oxygen-derived free radicals that target several molecules, including DNA and collagen [22]. During embryonic development, the fetus might also be exposed to a high-glucose environment during contexts such as diabetic embryopathy, which can cause excessive damage to multiple organs (e.g. the central nervous system and cardiovascular system) [22]. In addition, increased glucose availability can induce altered levels of prostaglandins and DNA biosynthesis, as well as differential expression of morphogenetic regulatory genes [22]. Thus, DNA damage can be induced by reactive oxygen species, resulting in the production of advanced glycation end products (AGEs) [23] that can lead to oxidative stress mediated by AGE receptors. This is observed in different organisms and is associated with several congenital abnormalities, such as genitourinary defects [12, 22]. Thus, metabolic perturbations can lead to cumulative damage in several tissues, such as blood vessels, nerves, muscles and eyes, which alters metabolic capacity and can be responsible for postnatal metabolic disorders, such as insulin resistance, diabetes mellitus and obesity [21, 22].

As discussed previously, our network presented proteins that are related to aerobic respiration. During embryogenesis, the establishment of the maternal-fetal interface facilitates vascular connections among placental tissues and the embryo, promoting an environment with an ambient O_2 supply that was previously hypoxic. This phenomenon induces angiogenesis and a gradually increasing O_2 supply.

The effects of low O_2 concentrations have been shown to promote the development of the morula and blastocyst when O_2 oscillations can modulate the gene expression of different lineages [24]. Additionally, appropriate levels of O_2 and aerobic metabolism can regulate a set of proteins that are related to epigenetic mechanisms, as well as the progression of cell fate decisions and cell cycle [12]. Thus, perturbations in aerobic respiration-related genes can promote prolonged hypoxic conditions, which can alter epigenetic mechanisms.

These relationships are consistent with the DevAge and DOHaD hypotheses, linking one more common regulatory network that begins during embryogenesis and is related to different phenotypes in aged individuals.

Summary of the Observed Mechanisms and Further Considerations

In this systems biology analysis, the major biological processes identified were immune system/inflammation, development, epigenetics and aerobic respiration/glucose metabolism. These processes are modulated during early development and are stringently subjected to selective pressure, which ensures successful development [12] (fig. 1). Richardson in 1999 [20] proposed the role of natural selection during various stages of development, which can act on regulatory networks and regulate adult morphologies. Thus, selective pressure is more predominant in the context of developmental mechanisms and adult trait specification, and small changes in either that occur during early development might result in significant morphological alterations during later stages [20].

In addition, the establishment of the maternal-fetal interface is dictated by the developmental conditions (e.g. mother's health and habits). Studies have shown that newborns are subject to maternal under- or overnutrition, and smoking during pregnancy has been implicated in morphological anomalies [19, 25]. Epigenetic programming is responsive to perturbations or imbalances of intrinsic and/or extrinsic factors experienced in utero. This can promote an adaptive response and influence gene expression patterns, leading to age-associated diseases, such as cancer, osteoporosis and the decline of the immune system [26]. In addition, noncoding RNAs have been implicated in chromosomal dynamics, highlighting its important role in the epigenetic regulatory control that has been implicated in developmental stages and cellular differentiation [27]. Studies have shown oscillations of miRNA expression in the brain during development, as well as during neurodegenerative conditions and age-associated disease, such as Alzheimer's disease. For example, the miRNA miR-107 is involved in the progression of Alzheimer's disease, but is underexpressed during the early pathogenesis of Alzheimer's disease [26].

In summary, aging is a programmed mechanism that is driven by a multifactorial developmental process. During development, these processes are crucial for the establishment of the placental environment, and they further culminate in tissue differentiation and embryonic survival. To overcome environmental challenges, the embryo needs to adapt its metabolism in response to environmental fluctuations, and such adaptations appear to be closely related to glucose metabolism, O_2 supply, the activation of a broad range of pro- and anti-inflammatory cytokines, and epigenetic modifications in chromatin structure.

It is possible to conclude that all of the hypotheses that link development and aging or that attempt to explain aging itself (e.g. DOHaD, TPH, IT, and antagonist pleiotropic theory) can all be linked to the DevAge theory. Lastly, our interactomic analysis corroborates the DevAge theory and aids in the understanding of the molecular mechanisms underlying aging and development.

References

1 Williams GC: Pleiotropy, natural selection, and the evolution of senescence. Evolution 1957;11:398–411.

2 Leroi AM, Bartke A, De Benedictis G, Franceschi C, Gartner A, et al: What evidence is there for the existence of individual genes with antagonistic pleiotropic effects? Mech Aging Dev 2005;126:421–429.

3 de Magalhães JP, Church GM: Genomes optimize reproduction: aging as a consequence of the developmental program. Physiology (Bethesda) 2005;20: 252–259.

4 Langley-Evans SC: Developmental programming of health and disease. Proc Nutr Soc 2006;65:97–105.

5 Hales CN, Barker DJP: Type 2 (non-insulin-dependent) diabetes mellitus: the thrifty phenotype hypothesis. Diabetologia 1992;35:595–601.

6 Hales CN, Barker DJ: The thrifty phenotype hypothesis. Br Med Bull 2001;60:5–20.

7 Dilman VM: Age-associated elevation of hypothalamic threshold to feedback control, and its role in development, aging, and disease. Lancet 1971;297:1211–1219.

8 Newman MEJ: The structure and function of complex networks. SIAM Rev 2003;45:167–256.

9 Scardoni G, Laudanna C: Centralities based analysis of complex networks; in Zhang Y (ed): New Frontiers in Graph Theory. http://www.intechopen.com/books/new-frontiers-in-graph-theory/centralities-based-analysis-of-networks.

10 Yu H, Kim PM, Sprecher E, Trifonov V, Gerstein M: The importance of bottlenecks in protein networks: correlation with gene essentiality and expression dynamics. PLoS Comput Biol 2007;3:e59.

11 Franceschi C, Bonafè M, Valensin S, Olivieri F, De Luca M, et al: Inflamm-aging. An evolutionary perspective on immunosenescence. Ann N Y Acad Sci 2000;908:244–254.

12 Feltes BC, de Faria Poloni J, Bonatto D: The developmental aging and origins of health and disease hypotheses explained by different protein networks. Biogerontology 2011;12:293–308.

13 Aller MA, Blanco-Rivero J, Arias JI, Balfagon G, Arias J: The wound-healing response and upregulated embryonic mechanisms: brothers-in-arms forever. Exp Dermatol 2012;21:497–503.

14 Barker DJ, Eriksson JG, Forsén T, Osmond C: Fetal origins of adult disease: strength of effects and biological basis. Int J Epidemiol 2002;31:1235–1239.

15 Wells JC: The thrifty phenotype: an adaptation in growth or metabolism? Am J Hum Biol 2001;23:65–75.

16 Bateson P, Gluckman P: Plasticity, Robustness, Development and Evolution. Cambridge, Cambridge University Press, 2011.

17 Wells JC: A critical appraisal of the predictive adaptive response hypothesis. Int J Epidemiol 2012;41:229–235.

18 Hochberg Z, Feil R, Constancia M, Fraga M, Junien C, Carel JC, et al: Child health, developmental plasticity, and epigenetic programming. Endocr Rev 2005;32:159–224.

19 Kajantie E: Early-life events. Effects on aging. Hormones (Athens) 2008;7:101–113.

20 Richardson MK: Vertebrate evolution: the developmental origins of adult variation. Bioessays 1999;21:604–613.

21 Hay WW Jr: Recent observations on the regulation of fetal metabolism by glucose. J Physiol 2006;572:17–24.

22 Kanwar YS, Nayak B, Lin S, Akagi S, Xie P, et al: Hyperglycemia: its imminent effects on mammalian nephrogenesis. Pediatr Nephrol 2005;20:858–866.

23 Barea F, Bonatto D: Relationships among carbohydrate intermediate metabolites and DNA damage and repair in yeast from a systems biology perspective. Mutat Res 2008;642:43–56.

24 Prado-Lopez S, Conesa A, Armiñán A, Martínez-Losa M, Escobedo-Lucea C, et al: Hypoxia promotes efficient differentiation of human embryonic stem cells to functional endothelium. Stem Cells 2010;28:407–418.

25 Feltes BC, de Faria Poloni J, Notari DL, Bonatto D: Toxicological effects of the different substances in tobacco smoke on human embryonic development by a systems chemo-biology approach. PLoS One 2013;8:e61743.

26 Munoz-Najar U, Sedivy JM: Epigenetic control of aging. Antioxid Redox Signal 2011;14:241–259.

27 Bernstein E, Allis CD: RNA meets chromatin. Genes Dev 2005;19:1635–1655.

Prof. Dr. Diego Bonatto
Centro de Biotecnologia da UFRGS – Sala 219
Departamento de Biologia Molecular e Biotecnologia
Universidade Federal do Rio Grande do Sul – UFRGS
Avenida Bento Goncalves 9500 – Predio 43421
Caixa Postal 15005
Porto Alegre – Rio Grande do Sul (Brazil)
E-Mail diegobonatto@gmail.com

Yashin AI, Jazwinski SM (eds): Aging and Health – A Systems Biology Perspective.
Interdiscipl Top Gerontol. Basel, Karger, 2015, vol 40, pp 85–98 (DOI: 10.1159/000364933)

Aging as a Process of Deficit Accumulation: Its Utility and Origin

Arnold Mitnitski[a, b] · Kenneth Rockwood[a, c]

Departments of [a]Medicine and [b]Mathematics and Statistics, Dalhousie University, [c]Division of Geriatric Medicine, QEII Health Science Centre, Halifax, N.S., Canada

Abstract

Individuals of the same age differ greatly with respect to their health status and life span. We have suggested that the health status of individuals can be represented by the number of health deficits that they accumulate during their life. We have suggested that this can be measured by a fitness-frailty index (or just a frailty index), which is the ratio of the deficits present in a person to the total number of deficits considered (e.g. available in a given database or experimental procedure). Further, we have proposed that the frailty index represents the biological age of the individual, and suggested an algorithm for its estimation. In investigations by many groups, the frailty index has shown reproducible properties such as: age-specific, nonlinear increase, higher values in women, strong association with mortality and other adverse outcomes, and universal limit to its increase. At the level of individual, the frailty index shows complex stochastic dynamics, reflecting both stochasticity of the environment and the ability to recover from various illnesses. Most recently, we have proposed that the origin of deficit accumulation lies in the interaction between the environment, the organism and its ability to recover. We apply a stochastic dynamics framework to illustrate that the average recovery time increases with age, mimicking the age-associated increase in deficit accumulation.   © 2015 S. Karger AG, Basel

People age at different rates. Even so, everyone ages, and to an end: the mortality rate is characteristic, rising exponentially with age. This is the case even though not everyone dies from the same illness. Despite the remarkable progress that has been made by studying how each individual illness arises, and how each contributes to mortality, there remains a strong rationale to consider illnesses in their totality. Multiple aging processes can be studied with profit by employing a systems biology approach [1–3]. Mathematical modeling, at the heart of systems biology, allows the useful insights and apparatus developed in other scientific fields to be brought to bear on understanding ageing. Here, we discuss how ageing can be understood as a process of deficit accumulation, which likely scales from the molecular level to produce macroscopically visible deficits [4]. After briefly reviewing major facts about the deficit accumulation approach to quantifying aging, we discuss a very general stochastic framework to explain how the origin of deficit accumulation stems from the interaction between the environment and the organism. Because the deficit accumulation approach to aging has many reproducible characteristics, we suggest that the biological age of an individual can be estimated in relation to how many deficits they have accumulated, compared with how many are accumulated, on average at their chronological age. We also discuss how this approach can be applicable in understanding differences in health status not only in individuals, but also between countries with differing socioeconomic status.

The Accumulation of Deficits as a Proxy Measure of Aging

However aging is defined, it should be measured. The simplest measure of aging is chronological age, not least because mortality rate increases exponentially with age, in accordance with the Gompertz law of mortality [5]. Even so, the mortality rate is a population characteristic, and its application to individuals is quite limited. That is because individuals at the same age differ from each other not only because of differences in their genetic profile, but also reflecting how the external environment affects the multiple biological pathways of damage control and repair. This variability in damage potential and control leads to variable changes in physiological functions, loss of homeostatic regulation, shrinkage of the homeodynamic space, and flexibility, and vulnerability to stressors [6]. In these ways, subcellular processes scale up to affect life expectancy and health span. This variability in the vulnerability to adverse outcomes (worsening health, mortality) is often referred to as *frailty* [7, 8]. In short, different people age at different rates, making the idea of biological age (which is intrinsically individual) quite attractive; even so, how to estimate biological age remains a matter of dispute. Despite great individual variability in how aging manifests, there is a universal characteristic – the number of health problems increases with age [9, 10]. In 2001, we suggested a method for appraising health status in older adults [9] later extended to younger ages [11]. We demonstrated how to use a simple count of deficits (symptoms, signs, functional impairment, and laboratory abnormalities), and in order to make the deficit count comparable across different datasets, we defined the *frailty index* as a ratio of the number of deficits that individuals accumulated to the total number of deficits available in the database. Remarkably, from this simple measure, it is possible to predict mortality and other adverse outcomes [7]. Since then, our group and many others have investigated the properties of the frailty index in multiple databases (~300,000 cases) from different countries including Canada, the United States, Australia, Mexico, China and most European countries. Together, the results are similar across databases that use different data collection methods (e.g. clinical vs. self-reported data), different designs (e.g. cross-sectional vs. longitudinal) and different numbers of variables. Specifically, the frailty index robustly shows that: (a) the average rate of deficit accumulation has a narrow range (in community-dwelling people between 2.5 and 4% per year, on a log scale) [7]; (b) men have lower mean frailty index values than do women of the same age; (c) women show better mean survival than do men who have the same frailty index value; (d) there is an empirically determined limit to the frailty index value close to 0.7; (e) the frailty index is a strong predictor of adverse outcomes; (f) the frailty index outperforms chronological age in all survival models.

The statistical distribution of the frailty index for older adults shows typically skewed pattern (fitted by the gamma density function). Of some note, people from institutions or with severe illnesses show bell-shaped patterns approximated by the normal distribution [9]. The frailty index can be regarded as a *clinical state variable*, in that it characterizes the whole health of individuals [7] and validly classifies risk across a wide range of people. In short, the frailty index, although nonspecific (but note, aging itself is *nonspecific*), is a very sensitive measure of health in older adults.

Remarkably, the values and age trajectories of frailty indices are not sensitive to the exact nature of the variables that might be included in the frailty index. Indeed, when some dozens or more variables are included, individual variables can be selected at random and still yield comparable results [12]. For this reason, and usefully for epidemiological inquires, the frailty index can be created from virtually any set of health-related variables if they cover a range of systems, are associated with adverse outcomes and increase with age [13]. The reason that the frailty index characterizes individuals' health independently of the variables it comprises can be understood if we recognize that the variables (health deficits) are not independent, even though this typically is assumed in the various risk factor models. Indeed, whatever statistical techniques might be employed to suggest otherwise, in any pragmatic sense the deficits are predominantly interdependent; each of them is linked with many others, i.e. each of them contains information about many others [9, 14].

One property of the frailty index is of considerable practical and theoretical interest. Each new deficit decreases the probability of surviving. There is an empirical limit to the frailty index around 0.7, beyond which further deficit accumulation is incompatible with life independently on their age [15]. A frailty index can also be constructed from clinical data using a Comprehensive Geriatric Assessment (CGA) from the items available in typical CGA forms [14]. This so-called FI-CGA has the usual properties of other versions of the frailty index – i.e. it is correlated with age, shows a skewed distribution, is higher in women, and correlates with several adverse outcomes, including institutionalization and health care use. It could aid clinical decision-making by indicating the *degree of frailty*, and thus the likelihood of an adverse outcome.

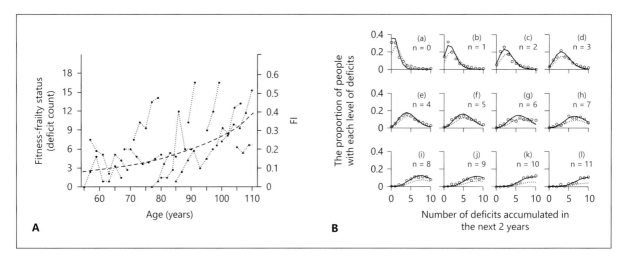

Fig. 1. A Examples of individual trajectories of the deficit count in 12 randomly selected participants of the NPHS. Individual trajectories are represented by dots connected by dashed lines. The age-specific deficit count (the average cross-sectional trajectory) is shown by the dashed line. Of note, the values of the individual numbers of deficits were changed to satisfy Statistics Canada privacy requirements, but the patterns have been preserved. From Mitnitski et al. [30]. **B** The probability of transition (y-axis) from a given deficit state *n* (shown in each subplot cell) to *k* deficits (x-axis). Circles represent observed transitions between two consecutive cycles. The lines show the model's fit according to equation 1. The solid line is for 2-year transitions. The data are truncated at 12 deficits for presentation; fewer than 5% of people show 12 or more deficits. From Mitnitski et al. [31].

Individual Frailty Trajectories: Stochastic Patterns

On average, health tends to decline with age, but in individuals, including older adults, health status can improve. In confirmation of such behavior, we have found that the individual values of frailty index do not monotonically increase over time; their changes are highly dynamic such that individuals may show significant change from a higher frailty index value (higher number of deficits) to a lower value reflecting a possibility of health improvement [14, 16]. This is illustrated in figure 1A, where individual courses typically show highly irregular patterns, including a few cases of 'big jumps' against a background of relatively small changes. Individual trajectories showing irregular behavior mark a stochastic process of changes in frailty states in relation to aging. Although highly irregular at the level of individuals, the process is nevertheless summarizable at a group level (fig. 1B). To reflect the complex nature of these changes, we introduced a stochastic model of health transitions between two fixed time points [16].

According to that model, the probability of transition from *n* to *k* deficits for the fixed time interval can be approximated by the following equation [16]:

$$P_{nk} = \frac{\rho_n^k}{k!} \exp\left(-\rho_n\right)\left(1 - P_{nd}\right). \tag{1}$$

Mitnitski · Rockwood

where P_{nd} is the probability of death and ρ_n is the Poisson mean which is state, n, dependent. It is usually well represented by the linear function on n: $\rho_n = \beta_0 + \beta_1 n$, and the probability of death is well represented by the logistic function, $logit(P_{nd}) = \alpha_0 + \alpha_1 n$.

In many settings and with many different measures, this model fits the observed data with very high precision as can also be seen in figure 1B, with the goodness of fit typically accounting for more than 90% of 'explained' variability. The major results are: (a) the model allows changes in health to be represented in all directions (including worsening and improvement) and at any degree (both gradual and in 'jumps'); (b) the model considers these changes and death simultaneously as competing events; (c) the model has a few interpretable parameters that depend on baseline conditions (i.e. the number of deficits at baseline); (d) the model is applicable not only to general health status (represented by the number of deficits) but also to other functions such as cognition. Note that the model illustrates dynamic aspects of how aging occurs through the accumulation of deficits; dynamics arise because of the possibility of repair and recuperation. Note that there is often fluctuation, and that changes in frailty status occur in both directions, i.e. even though, on average, decline predominates, an important proportion of people experience other trajectories.

As noted, an individual's likelihood of changing his/her health status is largely conditioned by his/her health at baseline (i.e. at the current evaluation). Most transitions, in either direction, are gradual, so that going from having no deficits to having many deficits, for example, is not common (and same for vice versa). From both clinical and public health standpoints, it is important to note two factors. First, this is a probabilistic model: for any individual, the model shows the chance of a series of outcomes, not simply the relative risk of a given outcome. Second, and relatedly, the model allows for the possibility of the real world phenomenon and clinical/public health goal of improvement. In other words, the model shows not just deficit accumulation, but deficit diminution (i.e. improvement).

Health Changes in the Fittest (The Zero State of Frailty)

The stochastic transition model considered above has an interesting characteristic that has potential in describing population health. Among the four parameters in equation 1, two (α_0 and β_0) characterize those people who had no deficits at baseline. People with no baseline deficits are said to be in the 'zero state', $n = 0$. People can be in the zero state at any age, although the older they are, the less likely this is to be the case. An example of how people transit from a baseline state (n) to any follow-up state (k) is illustrated in figure 1B, where cell (a) represents transitions from the zero state. We have evaluated the outcomes of people in the zero state at baseline in relation to their so-called social vulnerability index. Even in Canada, with a universal health care system and average Western indices of economic inequality, the 5-year mortality of those with the highest social vulnerability was more than twice that of those with the lowest social vulnerability [17]. The transitions of the fittest (i.e. the average number of deficits that they accumulate) are quantitative estimates of these background effects, and thereby appear to be a means of quantifying the health hazard of a given environment.

The Origin of Deficit Accumulation – A Stochastic Framework

Characterizing aging as a process of deficit accumulation gives us an opportunity to represent aging in an individual by the changes in the number of deficits this individual has. If so, the question can be posed: what is the cause of such accumulation? We will show that the origin of deficit accumulation can be understood by considering a very general process of interaction of the environment, which causes damage, and the ability of the organism to sustain/repair the damage. This can be considered under the framework of stochastic processes. To illustrate this, we assume that the process of environmental challenges imposing stresses on the organism can be considered as a stochastic process, with average intensity (rate) λ. The average time interval between the consecutive stresses is thus $1/\lambda$. Such challenges will be of many different sorts, arising from individual exposures to perturbations in the weather, solar activity, pollution, stressful social events, disease outbreaks, etc. The interval between the challenges ($1/\lambda$) varies from minutes to months; most cannot be measured. Let R be the average time of recovery from such environmental stresses in people of the same chronological age. The ability of the organism to recover depends on the individual's genetic profile [18], health status, living conditions, and access to modern health care [19]. Most importantly, the time of recovery is age dependent, presumably reflecting subclinical (even microscopic) tissue, cellular and subcellular damage [4].

Queuing Theory and Little's Law

There is a structural similarity between the process described above and the processes governing the formation of a queue described in stochastic queuing theory. Queuing theory is a mathematical discipline that aims to explain how the length of a queue is related to the intensity of the stream of arrivals to the system, and to the service and waiting times [20], e.g. how long it takes to service the people in the queue. Queuing theory is widely employed across multiple applications in communications, computer architecture, operation management, etc. The statistical mechanics of the queuing system can be complex, and even in simplest cases is described by the Kolmogorov differential equations. Although the specific details of these equations depend on the assumptions of the model (e.g. single server or a network of servers; stationary or non-stationary arrivals, different priory schedules, etc.), as outlined above, there is a general and simple relationship between the average number of items in a queuing system (N), the average arrival rate (λ) and the average waiting time of an item in the system, R, known as Little's Law [21] (here we use the different notation from the original [21]):

$$N = \lambda R \tag{2}$$

By exact analogy, we suggest that the average number of deficits present in an individual (N) equals the rate of environmental stresses λ (the amount of damage arriving to the queue), multiplied by the average recovery time R (analogous to waiting time) [22].

Damage occurs at the different levels of the organism, including DNA, epigenetics, cells, and organ tissues: each of these levels has its own repair mechanism; the latter are not isolated from each other but organized in complex networks, so the elimination of damage does not go through one isolated system with a queue but through many such subsystems. In some circumstances, repeated damage may happen at the same place and create a 'local' queue. It could also be some other reasons for the queue. Independent of the detailed mechanisms of these processes, equation 2 imposes fundamental constraint on them.

Age-Related Deficit Accumulation Reflects the Increase of Age-Related Time of Recovery

Equation 2 states that the average recovery time is proportional to the average number of deficits that the individual has. The coefficient of proportionality (λ) which is the intensity of the environmental stresses is an average characteristic of the environment. Let us assume that during the life course this average intensity does not change. As we know, deficits accumulate with age. According to equation 2, this means that the average recovery time changes proportionally to the increase in the number of deficits, i.e. the kinetics of the deficit accumulation with age is the same as the kinetics of recovery time. It has been demonstrated that the kinetics of deficit accumulation can be fitted by an exponential function [7, 9, 22] with the exponent parameter typically close to 0.03. Figure 2A shows age-specific (cross-sectional) average trajectories for the 9 waves of the National Population Health Survey (NPHS) of Canadians aged 20+ years, over 16 years of follow-up, repeated every 2 years. There is about a 3-fold increase in the number of deficits during 3 decades after the age of retirement (65 years) and about an order of magnitude increase from early adulthood to 100 years. Given the exponential increase in the number of deficits with age, according to equation 4 we can say that the *average* recovery time increases exponentially with age, and with the same parameter k = 0.035 [22]:

$$dR/dt = kR \tag{3}$$

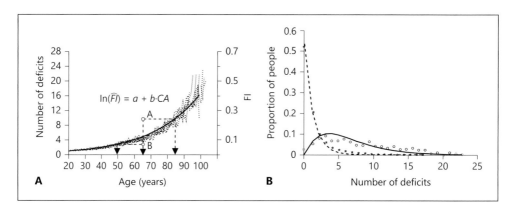

Fig. 2. A Age trajectories of the mean number of deficits. Thin lines are the cross-sectional data for the 9 consecutive 2-year cycles plotted against age. The solid line is the best exponential fit with the exponent of 0.035 ± 0.02. Two individuals A and B have the same chronological age of 65 years old: their biological ages, indicated by the arrows, correspond to those averages of the population represented by the calibration curve. From Mitnitski et al. [22]. **B** Distribution of the number of deficits for the two age groups: 20–30 years old (stars are the observational data and the dashed line is the exponential fit with the exponent –0.6) and 75–100 years old (circles are the observational data and the solid line is the gamma fit with the shape and scale parameters of 2.1 and 3.2, respectively). From Mitnitski et al. [22].

Even though the time it takes to recover from illness is age dependent [23, 24], the most severe conditions contribute to a greater extent to the average recovery time than relatively benign conditions.

Changes in the Distribution of the Number of Deficits Indicates Change in Recovery Potential

An important limitation of Little's Law is that it describes relationships only between the average characteristics. The stochastic dynamical model of deficit accumulation is needed to explain the changes in the distributions of the deficit accumulation over time. Note that these distributions change characteristically from being highly asymmetrical in younger people (most of whom have few deficits) to the gamma-like distributions seen in older people [22]. These patterns now can be understood from our model. When the recovery time is low, the damage caused by the environmental stresses is quickly repaired, which explains why most of young people have no or a few deficits. The length of the queue in a system with 'light traffic' shows the same highly asymmetrical density distributions as we see in younger people. As can be seen in figure 2B, in younger people (20–35 years old), the distribution of the number of deficits is highly asymmetrical (see a similar figure in Rockwood et al. [25]). The same figure also shows the density distribution of the number of deficits in an older group of individuals aged 75–100 years also fitted by the gamma density function typical for 'heavy traffic' models [22].

The age-related increase in recovery time can be related to the decreasing vitality first suggested in 1960 by Strehler and Mildvan [26]. The difference is that in their model vitality declines linearly while the rate of recovery changes exponentially. Here, too, increasing the recovery time with age appears also be related to a very general systems biology mechanism of critical slowing [27]. Finally, the origin of these changes can be related to the fact that the organism is not in equilibrium with the environment. The entropy of the organism is lower than the entropy of the environment, so that maintaining and keeping the structure of the organism requires permanent efforts for compensatory adaptation, repair and elimination of damage. The maintenance of such nonequilibrium causes increasing metabolic cost [2].

Estimating Biological Age

Although an important and desirable goal, the characterization of biological ageing has proven to be difficult, and for a long time was unsuccessful in tackling interindividual variability [1]. By long convention, the assessment of biological aging has translated into an often controversial search for *biomarkers of aging* – i.e. a search for specific nondisease traits which change over a life span. According to the usual terms of engagement in the debate, any biomarker should predict the outcome of a wide range of age-sensitive tests in multiple physiological domains in an age-coherent way, and do so better than chronological age. The many candidates considered as biomarkers of ageing have particularly included endocrine system/immune responses, those associated with caloric restriction, and cell membrane viscosity, the concentration of prostacyclin in fibroblasts, alterations of DNA methylation, compliance of the cardiovascular system and a host of several others. The chief challenge in finding any single biomarker is how to integrate across organ systems in a unified way so as to allow the estimation of biological age at the level of the whole organism. A standard response to this challenge is to evaluate many biomarkers using some form of multivariate regression analysis. Although common, this approach strikes us as deeply flawed. That is because as discussed above, candidate variables in such analyses, although held to be independent on statistical grounds can hardly be considered to be so: in biological systems, exactly the opposite generally holds. That is why results from any multivariate model typically do not prove to be generalizable; the output of these statistical exercises is contingent on the particular samples from which they are drawn. What should be evident, however, is that their application in other datasets will require recalculation of the weights which they employ; this level of precision is unlikely to be generalizable. Unless the model shows stable parameter estimates across databases, any theoretical understanding of biological age will be restricted.

The consistency of the relationship between the frailty index, age and health outcomes made it possible to suggest that it can be used as a basis for approximation of personal biological age [14, 28]. As an illustrative example (fig. 2A), consider 2 people, A and B, of the same chronologic age, for example, 65 years with their frailty index values deviated from the average. At 65 years, the mean value of the frailty index is 0.12. Person A has a frailty index value of 0.24, which is 0.12 higher than the mean value of the frailty index at 65 years (0.12), and rather corresponds to the mean value of the frailty index at age 80 years. In essence, person A has the health of an 85-years-old: although chronologically 65 years old, person A can be considered to be biologically 20 years older than chronologically. By contrast, person B has a frailty index value of 0.07, which is 0.05 lower than the mean frailty index value at age 50 years. In essence, person B has the health of a 50-year-old: although chronologically 65 years old, person B can be considered to be biologically 50 years old, i.e. 15 years younger than chronologically.

In other words, we used the relationships between the average frailty index and age as a calibration curve (fig. 2A).

$$\ln(\hat{FI}) = a + b \cdot CA \tag{4}$$

Where $\hat{FI}$ is the mean frailty index estimated as a function of age from the equation above that we can call a *calibration equation*. CA is the chronological age and a and b are empirical parameters that can be estimated from the data, using for example a least squares regression technique. Note that in general, the slope parameter b has been estimated in several databases to be about 0.03 [7]. For example, the most recent estimates obtained from the NPHS for people aged 15–105 years gave the estimates of $a = -4.16 \pm 0.08$ and $b = 0.035 \pm 0.001$ [22].

The biological age of an individual can be estimated from inversing equation 4, given the value of the FI of the individual:

$$BA = [\ln(FI) - a]/b \tag{5}$$

It can be seen that the difference between biological age and chronological age can be found from the following equation:

$$BA - CA = \ln(FI/\hat{FI})/b \tag{6}$$

In other words, biological age could be equal, greater or lower than chronological age. This approach therefore gives a ready metric for estimating by how many years an individual is younger or older than the average person of his/her chronological age in a given population. In that specific sense, people can be considered as biologically older compared to those who are in a better health who can be considered as biologically younger compared to those who are in worse health.

Equation 6 contains only one parameter, slope b, which has a narrow range of values [14]. For those individuals who had no reported deficits (i.e. the fittest people) biological age can be calculated by substituting FI = 0 by its next minimal value, which corresponds to one deficit present. The corresponding FI could be calculated accordingly: e.g. where 40 deficits are being considered, FI = 0.025, with 100 deficits FI = 0.01, etc.

Note too that the theoretical limit to BA (when FI = 1) is $- a/b$. For the NPHS data, max BA = 126 years. As maximally observed empirical limit of the FI = 0.7, the limit to BA according to equation 5 is 116 years, which is close to max life span observed in human.

The frailty index is fundamental to this method of calculating biological age; it is based on a simple count of deficits that are broadly defined, but biologically/clinically meaningful. The outcomes using predictive models based on this approach are highly generalizable – they typically show superior performance of the frailty index compared to chronological age [7, 11]. Biological age calculated based on the frailty index can become a useful means of assessing and monitoring health status in individuals.

Aging, Health, Wealth and Life Expectancy Worldwide

The numerous indicators of population health include health risk factors, disabilities, chronic disease and conditions, maternal and infant health, social determinants of health, etc. The deficit accumulation approach was evaluated in relation to national income and healthcare spending as well as their relationship with mortality. By assessing over 35,000 people from 15 European countries who participated in the Survey of Health Aging and Retirement in Europe (SHARE), we demonstrated that the frailty index constructed from 70 deficits showed significant inverse relationships with gross domestic product, health expenditures and mortality. Survival of frail people was higher in higher-income countries: the higher level of the frailty index generally corresponds to the lower level of gross domestic product [19]. Perhaps higher health expenditures in more developed countries can contribute to recovery from major illnesses and decrease the average recovery time. Similarly, the decrease in the intensity of the environmental stresses in countries with more developed social policies would also accord with our model with decreasing λ in equation 2. This is another large area of inquiry where the use of deficit accumulation approach can be promising.

Conclusions and Perspectives

Aging can be conceptualized as a process of the accumulation of deficits, taking place in different individuals in different ways, with a variety of rates for different organ systems. Deficit accumulation depends on the interplay of intrinsic and extrinsic factors. Deficits are indicators of physiological deregulation and therefore, by counting them, it is possible to quantify the level of such deregulation. This approach offers a simple and justifiable way of assessing health in individuals and populations.

Note that this approach to health deficits is silent on the nature of the deficit (e.g. disease, disability, symptoms, sign, laboratory or imaging or electrodiagnostic abnormality). The point merits further comments. First, it has proven to be surprisingly controversial, especially in the clinical literature. This is understandable for two reasons. Clinical training traditionally emphasizes diagnostic parsimony. A single cause for a large number of abnormalities is more likely to be correct than is an explanation that invokes multiple causes. (It is certainly more psychologically satisfying to the person who recognizes it.) Clinical training also emphasizes diagnostic precision: what works for condition A might be harmful for similarly looking condition B. (Confusion in a diabetic is a trivial example: giving insulin will help the patient whose confusion is due to blood sugar being too high, but be harmful to the diabetic in whom confusion reflects blood sugar being too low.) The disciple of precision and parsimony is not readily overcome, for good reason. Even so, it is less well suited to the reality of the nature of ageing, especially as the number of deficits increases. Attempts to 'disentangle disability and comorbidity for frailty' are rooted in this approach.

A second reason that the nature of the deficit being less important than the number of deficits has proven controversial is that at first glance it seems counter-intuitive. How can a skin problem and a heart attack be equivalent, in the sense of each simply being a single deficit? But the truth is that they can be. Not every heart attack is fatal. Not every rash is benign. To the extent that they impair function or induce a spiral of other diagnoses, they will add to the deficit count, and in that way the frailty index will capture their unequal nature in relation to prognosis.

Considering deficits in relation to age and not just diseases – i.e. adopting more a systems perspective – also appear to have implications for understanding the epidemiology of late life illness. An interesting report in this regard was published in *Neurology* in 2011, in the same issue as two other papers that reported 'novel risk factors for Alzheimer's disease' [29]. Instead, the third paper combined 19 so-called 'non-traditional dementia risk factors' in an index variable. The index variable (composed of items such as a history of diarrhea, dentures or foot problems) was a stronger risk for predicting all causes of late-life cognitive decline than was any traditional risk factor. The perspective offered here, that deficit accumulation reflects impaired recovery time, and that age-related recovery processes in the brain will not be unrelated to those in other parts of the body gives some broad insight into why deficits are so powerful, even if they are not known to be specific risk factors for the disease in question. Recent work by our group has replicated this observation in another cohort, and for both heart disease and osteoporosis.

To advance our understanding of age-related, multiply determined illnesses, the development of the statistical mechanics of deficit accumulations involving interaction between different subsystems of the human organism is the next natural step, and one that we are pursuing. This will allow better understanding of the processes of deficit accumulation, and their relationships with long- and short-term changes in the environment. This approach fits comfortably into overall agenda of extending the healthy life span.

References

1 Jazwinski SM: Biological aging research today: potential, peeves, and problems. Exp Gerontol 2002;37:1141–1146.

2 Kirkwood TB: Systems biology of ageing and longevity. Philos Trans R Soc Lond B Biol Sci 2011;366:64–70.

3 Yashin AI, Arbeev KG, Akushevich I, Kulminski A, Ukraintseva SV, Stallard E, Land KC: The quadratic hazard model for analyzing longitudinal data on aging, health, and the life span. Phys Life Rev 2012;9:177–188.

4 Howlett SE, Rockwood K: New horizons in frailty: ageing and the deficit scaling problem. Age Ageing 2013;42:416–423.

5 Gavrilov LA, Gavrilova NS: Reliability theory of aging and longevity; in Masoro EJ, Austad SN (eds): Handbook of the Biology of Aging, ed 6. Amsterdam, Academic Press, 2006, pp 3–42.

6 Rattan SI: Healthy ageing, but what is health? Biogerontology 2013;14:673–677.

7 Rockwood K, Mitnitski A: Frailty defined by deficit accumulation and geriatric medicine defined by frailty. Clin Geriatr Med 2011;27:17–26.

8 Clegg A, Young J, Iliffe S, Rikkert MO, Rockwood K: Frailty in elderly people. Lancet 2013;381:752–762.

9 Mitnitski AB, Mogilner AJ, Rockwood K: Accumulation of deficits as a proxy measure of aging. Sci World J 2001;1:323–336.

10 Yashin AI, Arbeev KG, Kulminski A, Akushevich I, Akushevich L, Ukraintseva SV: Cumulative index of elderly disorders and its dynamic contribution to mortality and longevity. Rejuvenation Res 2007;10:75–86.

11 Rockwood K, Mitnitski A, Song X: Changes in relative fitness and frailty across the adult lifespan: evidence from the Canadian National Population Health Survey. CMAJ 2011;59:814–821.

12 Rockwood K, Mitnitski A, Song X, Steen B, Skoog I: Long-term risks of death and institutionalization of elderly people in relation to deficit accumulation at age 70. J Am Geriatr Soc 2006;54:975–979.

13 Searle SD, Mitnitski A, Gahbauer EA, Gill TM, Rockwood K: A standard procedure for creating a frailty index. BMC Geriatr 2008;8:24.

14 Rockwood K, Mitnitski A: A clinico-mathematical model of aging; in Fillt HM, Rockwood K, Woodhouse K (eds): Brocklehurst'a Textbook of Geriatric Medicine and Gerontology, ed 7. Amsterdam, Elsevier, 2009, pp 59–65.

15 Rockwood K, Mitnitski A: Limits to deficit accumulation in elderly people. Mech Ageing Dev 2006;127:494–496.

16 Mitnitski A, Bao L, Rockwood K: Going from bad to worse: a stochastic model of transitions in deficit accumulation, in relation to mortality. Mech Ageing Dev 2006;127:490–493.

17 Andrew MK, Mitnitski A, Kirkland SA, Rockwood K: The impact of social vulnerability on the survival of the fittest older adults. Age Ageing 2012;41:161–165.

18 Yashin AI, Arbeev KG, Wu D, Arbeeva LS, Kulminski A, Akushevich I, Culminskaya I, Stallard E, Ukraintseva SV: How lifespan associated genes modulate aging changes: lessons from analysis of longitudinal data. Front Genet 2013;4:3.

19 Theou O, Brothers TD, Rockwood MR, Haardt D, Mitnitski A, Rockwood K: Exploring the relationship between national economic indicators and relative fitness and frailty in middle-aged and older Europeans. Age Ageing 2013;42:614–619.

20 Gross D, Harris CM: Fundamentals of Queuing Theory. New York, Wiley, 1998.

21 Little JDC: Little's law as viewed on its 50th anniversary. Oper Res 2011;59:536–549.

22 Mitnitski A, Song X, Rockwood K: Assessing biological aging: the origin of deficit accumulation. Biogerontology 2013;14:709–717.

23 Yanai H, Budovsky A, Tacutu R, Fraifeld VE: Is rate of skin wound healing associated with aging or longevity phenotype? Biogerontology 2011;12:591–597.

24 Akushevich I, Kravchenko J, Ukraintseva S, Arbeev K, Yashin AI: Recovery and survival from aging-associated diseases. Exp Gerontol 2013;48:824–830.

25 Rockwood K, Mogilner A, Mitnitski A: Changes with age in the distribution of a frailty index. Mech Ageing Dev 2004;125:517–519.

26 Strehler BL, Mildvan AS: General theory of mortality and aging. Science 1960;132:14–21.

27 Veraart AJ, Faassen EJ, Dakos V, van Nes EH, Lürling M, Scheffer M: Recovery rates reflect distance to a tipping point in a living system. Nature 2011;481: 357–359.

28 Mitnitski A, Rockwood K: Biological age revisited. J Gerontol A Biol Sci Med Sci 2014;69:295–296.

29 Song X, Mitnitski A, Rockwood K: Nontraditional risk factors combine to predict Alzheimer disease and dementia. Neurology 2011;77:227–234.

30 Mitnitski A, Song X, Rockwood K: Trajectories of changes over twelve years in the health status of Canadians from late middle age. Exp Gerontol 2012; 47:893–899.

31 Mitnitski A, Song X, Rockwood K: Improvement and decline in health status from late middle age: modeling age-related changes in deficit accumulation. Exp Gerontol 2007;42:1109–1115.

Arnold Mitnitski
Department of Medicine, Dalhousie University
Suite 1305, 5955 Veterans' Memorial Lane
Halifax, NS B3H 2E1 (Canada)
E-Mail Arnold.Miitnitski@dal.ca

Yashin AI, Jazwinski SM (eds): Aging and Health – A Systems Biology Perspective.
Interdiscipl Top Gerontol. Basel, Karger, 2015, vol 40, pp 99–106 (DOI: 10.1159/000364934)

Low-Grade Systemic Inflammation Connects Aging, Metabolic Syndrome and Cardiovascular Disease

Verónica Guarner · Maria Esther Rubio-Ruiz

Department of Physiology, Instituto Nacional de Cardiologia 'Ignacio Chávez', Mexico, Mexico

Abstract

Aging is associated with immunosenescence and accompanied by a chronic inflammatory state which contríbutes to metabolic syndrome, diabetes and their cardiovascular consequences. Risk factors for cardiovascular diseases (CVDs) and diabetes overlap, leading to the hypothesis that both share an inflammatory basis. Obesity is increased in the elderly population, and adipose tissue induces a state of systemic inflammation partially induced by adipokines. The liver plays a pivotal role in the metabolism of nutrients and exhibits alterations in the expression of genes associated with inflammation, cellular stress and fibrosis. Hepatic steatosis and its related inflammatory state (steatohepatitis) are the main hepatic complications of obesity and metabolic diseases. Aging-linked declines in expression and activity of endoplasmic reticulum molecular chaperones and folding enzymes compromise proper protein folding and the adaptive response of the unfolded protein response. These changes predispose aged individuals to CVDs. CVDs and endothelial dysfunction are characterized by a chronic alteration of inflammatory function and markers of inflammation and the innate immune response, including C-reactive protein, interleukin-6, TNF-α, and several cell adhesion molecules are linked to the occurrence of myocardial infarction and stroke in healthy elderly populations and patients with metabolic diseases.   © 2015 S. Karger AG, Basel

Aging is defined as a series of morphological and functional changes which take place over time. The term also refers to the deterioration of biological functions after an organism has attained its maximum reproductive potential [1, 2]. Aging is accompanied by a chronic inflammatory state which may contribute to metabolic syndrome (MS) and diabetes and their cardiovascular consequences. Inflammation accompanied by a proinflammatory cytokine production during aging is associated with predisposing factors that include increased oxidative stress, a decrease in ovarian function, a decrease in stress-induced glucocorticoid sensitivity and an increased incidence of asymptomatic bacteriuria. Indeed, when compared with young subjects, healthy elders are more stressed and show activation of the hypothalamus-hypophysis-adrenal axis [3].

MS is a number of criteria reflecting abnormalities in lipid and glucose metabolism. These abnormalities are considered to be a cause for atherosclerosis, cardiovascular disease (CVD) and type 2 diabetes mellitus. The prevalence of CVD among patients with diabetes is 3- to 5-fold higher than in patients without it. MS demonstrates ethnic and gender variants, its frequency depends on the lifestyle and age. MS in an elderly population is a proven risk factor for cardiovascular morbidity, especially stroke and coronary heart disease and mortality. The high prevalence of MS, heart attacks and diabetes in the elderly population evidences that age is an independent risk factor for the development of metabolic abnormalities [4].

CVDs appear as a consequence of both insulin resistance and inflammatory responses which are increased during aging. Risk factors for atherosclerosis and diabetes overlap, and there is a propensity of diabetic patients to have premature atherosclerosis leading to the hypothesis that both share an inflammatory and perhaps genetic basis [5]. Low-grade inflammation caused by the secretion by adipocytes of proinflammatory cytokines due to our thrifty genotypes and alterations in the innate immune system due to our proinflammatory genotype are linked to insulin resistance, diabetes and CVD [6–8].

Aging, Diseases and the Regulation of Energy Allocation

Stress response genes and nutrient sensors regulate energy directed to cell protection, maintenance and longevity; when food is plentiful and stress levels are low, genes support growth and reproduction; in contrast, harsh conditions favor a shift in gene activity towards cell protection and maintenance extending life span. Therefore, changes in diet that lead to obesity, MS and diabetes determine longevity and alter the aging process. Important genes in extending life span include kinase mammalian target of rapamycin, AMP-activated protein kinase, sirtuins and insulin/insulin like growth factor 1 (IGF-1) signaling. These genes integrate longevity pathways and metabolic signals in a complex interplay in which life span appears to be strictly dependent on substrate and energy bioavailability [9].

IGF-1-mediated signaling is determining for longevity. Abnormalities in the insulin signaling pathway generate age-related diseases and increased mortality, whereas the growth hormone/IGF-1 axis could potentially modulate longevity in many species. Moreover in humans, an age-related decline in IGF-1 levels occurs, and at old age, low IGF-1 levels are associated with frailty, poor nutrition and cognitive decline and an increased risk of death [10, 11].

The aging process is altered or accelerated when inflammation increases the propensity of metabolic diseases and CVD and the risk of diseases increases with age.

Aging and the Immune System

Aging has been associated with immunological changes, denominated immunosenescence. An elderly immune system becomes more and more predisposed to chronic inflammatory reactions and is less able to respond to acute and massive challenges by new antigens. A young immune system has to cope quickly and efficiently with acute immunological challenges to assure survival and the reaching of reproductive age. Such reaction capability gradually burns out because of lifelong antigenic attrition. Moreover, lifelong antigenic challenges and the increasing antigenic burden determine a condition of chronic inflammation, with increased lymphocyte activation and proinflammatory cytokines [12].

Polymorphisms in the promoter regions of pro- and anti-inflammatory cytokine genes influence the level of cytokine production and the aging process. Nutrients with anti-inflammatory properties, such as vitamin E and n-3 polyunsaturated fatty acid, may reduce the level of chronic inflammation and thereby ameliorate tissue and functional loss during aging. New evidence suggests that, for the latter nutrient, gene-nutrient interactions occur that alter the effectiveness of dietary therapy [13].

Inflammation in Obesity and Metabolic Syndrome during Aging

Obesity is increased in the elderly population. Obesity is the result of a complex interaction of factors in each individual including: genetic predisposition, diet, metabolism and physical activity. The increase in the mass of adipose tissue induces a state of systemic inflammation due to an increase in secretory factors (adipokines) derived from pre-adipocytes and from macrophages constituting this tissue (fig. 1). This inflammation significantly contributes to endothelial dysfunction present in the CVD developed as a consequence of MS and diabetes [14].

Adipose tissue also provides energy for the immune system, which has a significant energy cost. The contribution of energy stores to immune function became clear from early studies noting reduced survival in subjects of low relative weight [15, 16]. Infection imposes a metabolic burden on account of the need to synthesize immunoglobulins and acute-phase proteins and other processes such as inflammation and fever. To meet these costs of infection, lipolytic factors such as cortisol, glucagon and various hormones release energy from adipose tissue [17, 18].

The immune system represents a priority function of adipose tissue during malnutrition. Adipose tissue has been previously considered as a toxic substance, but it may

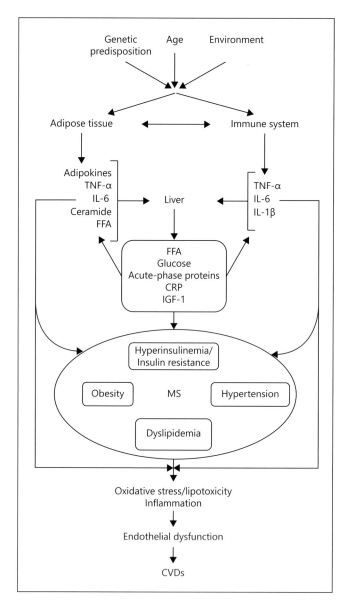

Fig. 1. Diagram of the pathophysiology of MS associated with low-grade systemic inflammation. Aging is accompanied by a chronic inflammatory state which may contribute to MS. The figure indicates the involvement of inflammatory factors derived from adipose tissue, liver and immune system cells leading to endothelial dysfunction and contributing to the development of CVDs. FFA = Free fatty acids.

be more appropriate to consider it as an activator of the immune function to increase protection against infectious diseases.

Adipocytes from old mice induce a higher inflammatory response in other cells. Sphingolipid ceramide is higher in old than in young adipocytes. Reducing ceramide levels or inhibiting NF-κB activation decreases cytokine production, whereas the addition of ceramide increases cytokine production in young adipocytes to a level comparable to that seen in old adipocytes, suggesting that ceramide-induced activation of NF-κB plays a key role in inflammation [19] (fig. 1).

The Immune System and Type 2 Diabetes

There is increasing evidence that an ongoing cytokine-induced acute-phase response is closely involved in the pathogenesis of type 2 diabetes mellitus and associated complications such as dyslipidemia and atherosclerosis. Elevated circulating inflammatory markers such as C-reactive protein (CRP) and interleukin-6 (IL-6) predict the development of type 2 diabetes mellitus, and several drugs with anti-inflammatory properties (aspirin and thiazolidinediones) lower both acute-phase reactants and glycemia and possibly decrease the risk of developing type 2 diabetes mellitus (statins). Among the risk factors for type 2 diabetes mellitus, which are also known to be associated with activated innate immunity, are age, inactivity, certain dietary components, smoking, psychological stress, and low birthweight. Other features of type 2 diabetes mellitus, such as fatigue, sleep disturbance and depression, are likely to be at least partly due to hypercytokinemia and activated innate immunity [20].

The Liver and Inflammation in Metabolic Syndrome

The liver plays a pivotal role in the metabolism of nutrients, drugs, hormones, and metabolic waste products, thereby maintaining body homeostasis. The liver is central to glucose and lipid homeostasis as well as steroid biosynthesis and degradation. This organ also has a major impact on health and homeostasis through its control of serum protein composition. Concomitant with morphological changes, the liver exhibits important alterations in global gene expression profiles with age. In mice, aging is accompanied by changes in expression of genes associated with increased inflammation, cellular stress, fibrosis, altered capacity for apoptosis, xenobiotic metabolism, normal cell-cycle control, and DNA replication. These changes predispose aged individuals to CVD [21].

Hepatic steatosis and its related inflammatory state (steatohepatitis) are the main hepatic complications of obesity and metabolic diseases. Hepatic steatosis is a disorder characterized by fat infiltration and excessive accumulation of lipids such as triglycerides in the liver (nonalcoholic fatty liver, NAFLD). The accumulation of fat in hepatocytes is a consequence of three principle sources: de novo lipogenesis in the liver; nutritional uptake from the small intestine; and free fatty acid release from visceral white adipose tissue.

Hepatic steatosis is accompanied by an increased liver/body weight ratio and higher plasma levels of enzyme markers of liver damage (alanine aminotransferase, γ-glutamyltransferase, and alkaline phosphatase). This pathology, which is often associated with obesity, hyperinsulinemia, and insulin resistance, shows an inflammatory state, characterized by increased hepatic and plasma levels of several proinflammatory cytokines, particularly TNF-α, which may play a crucial role in the progress of steatohepatitis to hepatic necrosis, fibrosis, cirrhosis and cancer (fig. 1).

It has been observed that 70% of the adult patients and 25.5% of the pediatric patients with MS have NAFLD. The prevalence of NAFLD increases with age, but the underlying molecular mechanisms need to be further investigated. Indeed, aged mice both under standard diet conditions or a high-fat diet will develop hepatic steatosis.

Hepatocytes, like other secretory cells, are rich in endoplasmic reticulum (ER). The ER is a highly dynamic organelle that has essential roles in multiple cellular processes that are required for cell survival and normal cellular functions. ER stress contributes to the pathology of many human diseases. Cell death, a physiological consequence of chronic ER stress, is key to the pathogenesis of many diseases including obesity, insulin resistance, hepatic steatosis, inflammation, neurodegenerative disorders and cancer.

The ER responds to environmental stress such as hyperlipidemia, hyperhomocysteinemia, hyperglycemia, and inflammatory cytokines, triggering a series of signaling cascades known as the unfolded protein response (UPR). The primary signal that activates the UPR is the accumulation of misfolded proteins in the ER lumen. As a consequence, the UPR regulates the size, shape and components of the ER to accommodate fluctuating demands on protein folding, as well as other ER functions in coordination with different physiological and pathological conditions. ER stress activates NF-κB and JNK, with downstream effects on inflammatory recruitment, phosphorylation of insulin receptor signaling intermediates (to worsen insulin resistance), lipogenesis, and oxidative stress. Hence, it is important to seek strategies to improve the antioxidant capacity in subjects who suffer from NAFLD as a consequence of MS.

Aging-linked declines in expression and activity of key ER molecular chaperones and folding enzymes compromise proper protein folding and the adaptive response of the UPR [22].

Fatty acids acting through toll-like receptors (TLR) in hepatocytes increase inflammation. TLR receptors are important pattern recognition receptors in the immune system that identify bacterial pathogens, but recently their participation in hypertension and insulin resistance has been recognized. Eight TLRs are expressed in mammalian liver (TLRs 1, 2, 4, 6–10). Individual TLRs interact with different combinations of adapter proteins and activate transcription factors such as NF-κB and JNK/activator protein 1. JNK activation is a key injury and inflammatory pathway in MS-related NAFLD [23].

Inflammatory Function, Atherosclerosis and Other Cardiovascular Consequences

Inflammation is one of the main mechanisms underlying endothelial dysfunction, and therefore it plays an important role in atherosclerosis and other CVDs such as hypertension. Recent investigations of atherosclerosis have focused on inflammation, providing new insight into mechanisms of the disease. Atherosclerosis is a disorder characterized by a chronic alteration of inflammatory function, and key markers of inflammation and the innate immune response, including CRP, IL-6, TNF-α, and several cell adhesion molecules are linked to the occurrence of myocardial infarction and stroke in both healthy populations and among those with known coronary disease [24] (fig. 1).

Inflammatory cytokines involved in vascular inflammation stimulate the generation of endothelial adhesion molecules, proteases, and other mediators, which may enter the circulation in soluble form. The concept of the involvement of inflammation in atherosclerosis has spurred the discovery and adoption of inflammatory biomarkers for cardiovascular risk prediction. CRP is currently the best validated inflammatory biomarker; in addition, soluble CD40 ligand, adiponectin, IL-18, and matrix metalloproteinase 9 may provide additional information for cardiovascular risk stratification and prediction.

An enhanced immune response also increases plaque vulnerability. Enhanced inflammation might prove to be an evolutionary determinant of atherogenesis, plaque rupture, platelet aggregation, and acute thrombosis.

Aging and hyperglycemia contribute to reduced mitochondrial biogenesis and mitochondrial dysfunction. These mitochondrial abnormalities can predispose a metabolic cardiomyopathy characterized by diastolic dysfunction. Mitochondrial dysfunction and resulting lipid accumulation in skeletal muscle, liver, and pancreas also impede insulin metabolic signaling and glucose metabolism, ultimately leading to a further increase in mitochondrial dysfunction [25].

Free oxygen radicals are involved in alcoholic cardiomyopathy, ischemia-reperfusion injury and aging. The myocardial cells are an important source of free radicals. When this organ suffers from diminished blood supply to an area as a result of diverse conditions such as a stroke, ischemia produces oxidative stress and structural damage, and the affected tissues die due to necrosis. Reperfusion may reverse the lethal process, but often not without taking its toll in the form of injury to the tissues. This is due to calcium re-entry to the cell, and this also generates an important amount of free radicals which are linked to alterations in mitochondrial function. There are specific alterations in heart mitochondrial function which occur as a result of ischemia and reperfusion and they involve the electron transport complexes, ATP concentration, ADP/ATP translocase, permeability transition and uncoupling [26].

Conclusion

Aging is associated with immunological changes, denominated immunosenescence, and is accompanied by a chronic inflammatory state which may contribute to MS and diabetes and their cardiovascular consequences. Inflammation is enhanced in the elderly population since there is increased obesity that increases fat-produced cytokines and alterations in hepatic function that lead to inflammation. Risk factors for CVDs and metabolic diseases overlap, and therefore the hypothesis that they share an inflammatory basis has been proposed, suggesting that low-grade systemic inflammation connects aging, MS and CVD.

References

1 Guarner V, Rubio-Ruiz ME: Metabolic syndrome: early development and aging. J Diabetes Metab DOI: 10.4172/2155-6156.S2-002.

2 Guarner-Lans V, Rubio-Ruiz ME: Aging, metabolic syndrome and the heart. Aging Dis 2012;3:269–279.

3 Bauer ME: Stress, glucocorticoids and ageing of the immune system. Stress 2005;8:69–83.

4 Tereshina EV: Metabolic abnormalities as a basis for age-dependent diseases and aging? State of the art. Adv Gerontol 2009;22:129–138.

5 Stern MP: Diabetes and cardiovascular disease: the common soil hypothesis. Diabetes 1995;44:369–374.

6 Pickup JC: Inflammation and activated innate immunity in the pathogenesis of type 2 diabetes. Diabetes Care 2004;27:813–823.

7 Festa A, D'Agostino R Jr, Howard G, Mykkänen L, Tracy RP, Haffner SM: Chronic subclinical inflammation as part of the insulin resistance syndrome: the Insulin Resistance Atherosclerosis Study (IRAS). Circulation 2000;102:42–47.

8 Kobayasi R, Akamine EH, Davel AP: Oxidative stress and inflammatory mediators contribute to endothelial dysfunction in high-fat diet induced obesity in mice. J Hypertens 2010;28:2111–2119.

9 Kenyon CJ: The genetics of aging. Nature 2010;464: 504–512.

10 Rincon M, Rudin E, Barzilai N: The insulin/IGF-1 signaling in mammals and its relevance to human longevity. Exp Gerontol 2005;40:873–877.

11 Rozing MP, Westendorp RGJ, Frölich M, de Craen AJM, Beekman M, Heijmans BT, Mooijaart SP, Blauw GJ, Slagboom PE, van Heemst D; Leiden Longevity Study (LLS) Group: Human insulin/IGF-1 and familial longevity at middle age. Aging 2009;1: 714–722.

12 Gersh BJ, Tsang TS, Seward JB: The changing epidemiology and natural history of nonvalvular atrial fibrillation: clinical implications. Trans Am Clin Climatol Assoc 2004;115:149–160.

13 Grimble RF: Inflammatory response in the elderly. Curr Opin Clin Nutr Metab Care 2003;6:21–29.

14 Frigolet ME, Torres N, Tovar AR: White adipose tissue as endocrine organ and its role in obesity. Arch Med Res 2008;39:715–728.

15 Lord G: Role of leptin in immunology. Nutr Rev 2002;60:S35–S38.

16 Kuzawa CW: Adipose tissue in human infancy and childhood: an evolutionary perspective. Am J Phys Anthropol 1998;(suppl 27):177–209.

17 Scrimshaw NS: Energy cost of communicable diseases in infancy and childhood; in Schurch B, Scrimshaw NS (eds): Activity, Energy Expenditure and Energy Requirements of Infants and Children. Vienna, IDECG, 1990, pp 215–237.

18 Biesel W: Metabolic response to infection. Ann Rev Med 1975;26:9–20.

19 Wu D, Ren Z, Pae M, Guo W, Cui X, Merrill AH, Meydani SN: Aging up-regulates expression of inflammatory mediators in mouse adipose tissue. J Immunol 2007;179:4829–4839.

20 Pickup JC: Inflammation and activated innate immunity in the pathogenesis of type 2 diabetes. Diabetes Care 2004;27:813–823.

21 Lebel M, de Souza-Pinto NC, Bohr VA: Metabolism, genomics, and DNA repair in the mouse aging liver. Curr Gerontol Geriatr Res DOI: 10.1155/2011/859415.

22 Brown MK, Naidoo N: The endoplasmic reticulum stress response in aging and age-related diseases. Front Physiol DOI: 10.3389/fphys.2012.00263.

23 Farrell GC, van Rooyen D, Gan L, Chitturi S: NASH is an inflammatory disorder: pathogenic, prognostic and therapeutic implications. Gut Liver 2012;6:149–171.

24 Packard RR, Libby P: Inflammation in atherosclerosis: from vascular biology to biomarker discovery and risk prediction. Clin Chem 2008;54:24–38.

25 Ren J, Pulakat L, Whaley-Connell A, Sowers JR: Mitochondrial biogenesis in the metabolic syndrome and cardiovascular disease. Mol Med (Berl) 2010;88: 993–1001.

26 Baños G, El Hafidi M, Franco M: Oxidative stress and cardiovascular physiopathology. Curr Topics Pharmacol 2000;5:1–17.

Verónica Guarner, PhD
Departamento de Fisiología, Instituto Nacional de Cardiología 'Ignacio Chávez'
Juan Badiano 1, Tlalpan
México, D.F. 14080 (México)
E-Mail gualanv@yahoo.com

Yashin AI, Jazwinski SM (eds): Aging and Health – A Systems Biology Perspective.
Interdiscipl Top Gerontol. Basel, Karger, 2015, vol 40, pp 107–127 (DOI: 10.1159/000364974)

Modulating mTOR in Aging and Health

Simon C. Johnson · Maya Sangesland · Matt Kaeberlein ·
Peter S. Rabinovitch

Department of Pathology, University of Washington, Seattle, Wash., USA

Abstract

The physiological responses to nutrient availability play a central role in aging and disease. Genetic and pharmacological studies have identified highly conserved cellular signaling pathways that influence aging by regulating the interface between nutrient and hormone cues and cellular growth and maintenance. Among these pathways, the mechanistic target of rapamycin (mTOR) has been most reproducibly shown to modulate aging in evolutionarily diverse organisms as reduction in mTOR activity extends life span from yeast to rodents. mTOR has been shown to play a role in a broad range of diseases, and is of particular interest to human health and aging due to the availability of clinically approved pharmacological agents targeting the mTOR complexes and other components of the mTOR signaling network. Characterizing the role of mTOR in aging and health promises to provide new avenues for intervention in human aging and disease through modulation of this signaling pathway.   © 2015 S. Karger AG, Basel

mTOR and Aging

The mechanistic target of rapamycin, mTOR, is a highly conserved serine/threonine kinase that plays a central role in sensing and responding to nutrient availability and growth signaling in eukaryotes. mTOR, encoded in mammals by *MTOR*, is an essential component of two distinct multiprotein complexes, mTORC1 and mTORC2. These signaling complexes regulate a variety of basic cellular activities including growth rate, cell size, and metabolism, and act as critical signaling hubs at the interface between nutrient or hormonal cues and cell growth and maintenance.

mTOR was first identified in yeast in studies of the immune-suppressive compound rapamycin. Rapamycin had previously been shown to act by forming a gain-of-function complex with Fpr1p, with mutations in *FPR1* conferring recessive resistance to rapamycin, and expression of the human homolog FKBP12 (FK-506 binding protein 12) restoring drug sensitivity. Mutations in *TOR1* and *TOR2*, originally designated *DDR1* and *DDR2* (dominant rapamycin resistance), were found to confer resistance to the antiproliferative effects of rapamycin in yeast. Wild-type Tor1 and Tor2 are bound and inhibited by the Fpr1/rapamycin complex [1]. This mechanism was subsequently found to be conserved in mammals with the rapamycin-FKBP12 complex binding to and inhibiting mTOR [2].

A link between mTOR signaling and aging was first established in yeast when studies in *Saccharomyces cerevisiae* demonstrated that deletion of Sch9, the yeast homolog of the mTORC1 substrate S6K (see 'mTOR Downstream Signaling'), results in a significant increase in chronological life span, defined as the duration of time that a yeast population retains viability when in a nondividing state [3]. Studies in the nematode *Caenorhabditis elegans* subsequently revealed that mTORC1 can negatively regulate longevity in multicellular organisms; knockdown of *daf-15* (the nematode homolog of Raptor, a component of mTORC1) or *let-363* (the nematode homolog of mTOR) by RNAi can extend life span in this model [4]. Reports from *Drosophila melanogaster* and yeast replicative aging studies further supported the role of mTOR in regulating longevity in lower eukaryotes.

Direct evidence of a role for mTOR in mammalian life span has been provided by studies showing life span extension in mice resulting from deletion of S6K, by double heterozygosity for mTOR and mlst8 (a component of mTORC1), and by treatment with rapamycin [5]. Intriguingly, life span extension in each of these studies was strongly sex-specific, with males receiving no longevity benefit from S6K deletion or double heterozygosity of mTORC1 components. Female mice also experienced a more robust response to treatment with rapamycin with an 18% increase in median life span, compared to a 10% increase in male animals [5]. Notably, intervention with rapamycin resulted in an increase in life span even when rapamycin treatment began late in life, suggesting that mTOR inhibition may prove an attractive target for intervening in human aging.

mTOR Signaling

mTOR in Nutrient and Growth Factor Sensing

mTORC1 and mTORC2 both play essential roles in eukaryotic biology, as complete loss of either Raptor, an mTORC1 specific component, or Rictor, an mTORC2 component, results in embryonic lethality. While both are essential for development, the two mTOR complexes differ in their components, relative regulatory roles, and upstream regulation of their activity. The upstream regulators and downstream effectors of mTORC1 are generally better characterized than those related to mTORC2. Although mTOR signaling is affected by numerous intra- and extracellular growth cues

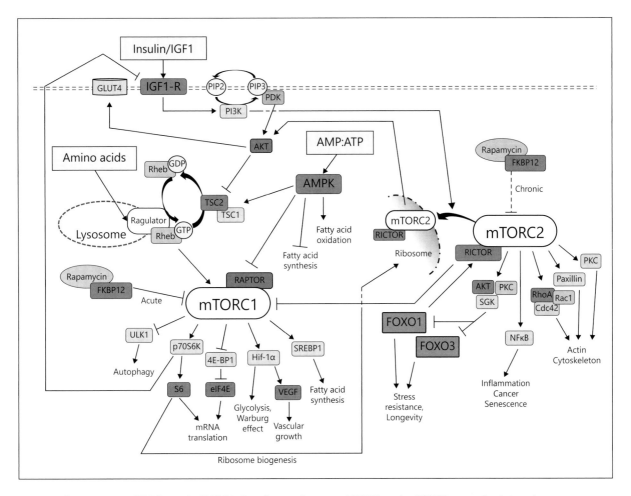

Fig. 1. Major mTORC1 and mTORC2 signaling pathways. mTORC1 and mTORC2 are major intracellular signaling hubs that respond to a variety of stimuli and mediate an array of downstream responses. Major regulators of mTORC1 signaling through insulin and IGF, sensing of intracellular amino acid levels at the lysosome via ragulator, and response to intracellular AMP:ATP levels through sensing by AMPK. mTORC2 is activated at assembled ribosomes and by Foxo1. Downstream pathways regulated by mTORC1 include mRNA translation, autophagy, metabolism, and vascularization. mTORC2 regulates actin/cytoskeletal assembly, inflammation, and stress resistance mediated by Foxo1 and Foxo3.

and conditions, and it affects numerous downstream pathways and processes, only the best characterized of these pathways, in terms of aging and disease, are described below.

Insulin/Insulin-Like Growth Factor/Phosphoinositide/AKT

mTOR is activated by a variety of growth factors and mitogens (fig. 1). Among the canonical regulators of mTORC1 signaling in mammals are insulin and the insulin-like growth factors (IGFs). Insulin and IGFs are recognized at the cell surface by tyrosine kinase receptors and provide the primary extracellular regulation of mTOR. Signaling through insulin/IGF is partially through phosphoinositide (PI3)-mediated activation PI3-dependent kinase (PDK) and subsequent activating phosphorylation of AKT on T308 [5]. IGF-1 signaling represents a longevity-regulating pathway in its own right, acting through both mTOR and through FoxO. IGF receptor loss has been shown to increase life span in mice and worms (*daf-12* is the IGF homolog in *C. elegans*); serum IGF levels have been shown to correlate with life span among mouse strains, and FOXO gene variants are strongly associated with extreme longevity in humans [6]. IGF-1 is the canonical and best characterized IGF activator of mTOR. AKT stimulates mTORC1 through phosphorylation of the mTORC1 inhibitor tuberous sclerosis complex protein 2 (TSC2). In its active form, the TSC1/2 complex is a GTPase-activating factor for the small guanine nucleotide-binding protein Rheb. Stimulation of Rheb by active TSC1/2 results in a conversion of loaded GTP to GDP, inactivating the protein. Active GTP-bound Rheb is a necessary component of mTORC1; thus, inhibition of TSC1/2 results in downstream activation of mTORC1 through an increase in active Rheb [7].

Ribosome Capacity

While mTORC2 is not activated by insulin and growth factors through canonical signaling events, an intriguing mechanism has been described which couples regulation of mTORC2 activity by extracellular signals to intracellular ribosomal capacity [8]. In this pathway, mTORC2 is activated by PI3K through increased physical association with ribosomes. Activated mTORC2 at the ribosome phosphorylates S473 on a hydrophobic motif of Akt, priming it for activation by PDK1. Ribosome-associated active mTORC2 also phosphorylates a hydrophobic motif of SGK and PKC. The relationship between mTORC2 and ribosome activity links cellular translation capacity to growth signaling. Simultaneously, ribosomal biosynthesis is also regulated by TORC1 activation of p70S6K and its target, the ribosomal S6 subunit. This crosstalk between mTORC2 and mTORC1 provides another link between intracellular growth capacity sensing and the regulation of growth by mTOR.

mTORC2 thus lies both upstream and downstream of AKT, a relationship that may explain a variety of otherwise inexplicable observations related to mTOR biology. One such case is the uncoupling of mTORC1 and AKT signaling by Foxo induction. Foxo has been found to upregulate levels of Rictor, the mTORC2-specific binding partner of mTOR, and activate AKT while simultaneously causing a decrease in assembled mTORC1. This decrease has been shown to result from sequestration of mTOR to mTORC2, while increased activity of mTORC2 directly activates AKT. The downstream effects of mTORC2-specific signaling are not well characterized in the context of aging but this complex relationship with mTORC1 and AKT demonstrates the importance of mTORC2 in the outcome of mTOR signaling perturbations.

AMPK

mTOR activity is modulated not only through extracellular signaling, but also by multiple intracellular energy and nutrient-sensing pathways. One well-characterized regulator of mTOR activity is the highly conserved AMP-activated kinase, AMPK. AMPK is an ancient sensor of energy status that acts as an intracellular upstream regulator of mTOR. AMPK is sensitive to the AMP:ATP status in the cell and is activated as this ratio increases. Upon activation, AMPK drives catabolic processes, such as fatty acid metabolism, and inhibits anabolic reactions, such as lipid synthesis, for the purpose of balancing intracellular energy status. AMPK acts on mTOR through multiple interactions. TSC2 is activated by AMPK through phosphorylation at T-1227 and S-1345. AMPK also appears to directly inhibit mTORC1 through phosphorylation of Raptor at serine 722 and serine 792 [9].

Multiple pharmaceutical agents that target AMPK signaling are available for research use, and AMPK activators have been used clinically in the setting of type 2 diabetes. Phenformin, a potent and direct AMPK activator, was used for the treatment of type 2 diabetics for decades before being removed from clinical use. This drug had strong beneficial effects for patients but was found to cause life-threatening, sometimes lethal, lactic acidosis following exercise [10]. The less potent and less characterized drug metformin replaced phenformin in the clinic. Metformin appears to have many of the same biological effects as phenformin (primarily inhibition of glycolysis and regulation of blood glucose) with a much lower frequency of the life-threatening lactic acidosis. The exact mechanism of function for metformin is not well understood, with some recent studies suggesting that the compound functions through inhibition of complex I of the electron transport chain, and others demonstrating an effect from the drug even in AMPK-deficient animals, together suggesting that metformin functions in large part through AMPK-independent pathways.

Amino Acid Sensing at the Lysosome

mTORC1 is also directly activated by amino acids through an interaction at the surface of the lysosome, allowing for intracellular amino acid abundance to regulate growth and metabolism directly through mTOR [11]. This sensing is facilitated by the recently discovered ragulator complex, a vacuolar ATPase binding complex that contains a guanine nucleotide exchange factor for Rheb and is regulated at the lysosomal surface by amino acids. Under conditions of high intracellular amino acids, mTORC1 and Rheb are recruited to the ragulator complex at the surface of the lysosome where Rheb activates mTORC1 as described above. Together, the regulatory interactions described above define mTORC1 as a central sensor of growth conditions and energy cues.

Caloric Restriction

Caloric restriction (CR) is a well-documented life span-extending intervention that has been found to be effective in divergent species including yeast, flies, rodents, and nonhuman primates [12]. Defined as a reduction in nutrient intake in the absence of malnutrition, CR (also referred to as dietary restriction or DR) is simple in theory but is a complicated intervention in practice, and one that has not been rigorously standardized. CR strategies differ greatly between organisms, life span assay type, laboratory, and field of study [13]. Even within the same organism, CR can be implemented in a variety of fashions. In nematodes, CR can be accomplished by titrating down the bacterial food available on solid plates or in media or by removing bacteria altogether (referred to as bacterial deprivation or BD). A variety of CR treatments exist in yeast including variable degrees of glucose deprivation or substitution of glucose (the standard carbon source for yeast on plates) with nonfermentable carbon sources. Murine studies of CR vary greatly in terms of food composition, percent CR (mass of food provided compared to ad libitum intake), and housing conditions, such as implementing CR on singly versus multiply housed mice.

Given that CR varies widely in implementation practices, it is almost surprising that CR generally works, suggesting that the downstream factors regulating this intervention are robust, well conserved, and may hold real promise in human aging and disease. There has been a significant effort to identify downstream effectors of CR, and mTOR has been often identified as a pathway likely to regulate a portion of the CR response. Given the role of mTOR in nutrient signaling and response to growth cues, it makes sense that mTOR would be involved in the CR response. mTORC1 activity has been shown to be reduced by CR in invertebrate and mammalian studies. Additional evidence implicating mTORC1 in CR comes from reports in which CR fails to additively extend yeast replicative life span in mutants for mTOR or S6K, as well as from studies in *C. elegans* showing a similar lack of additive extension between a genetic model for CR (the eat-2 mutant) and RNAi against mTOR [14]. S6K and translation initiation factors were, however, found to be additive with CR in this model.

It has become clear over the past decade of aging research in invertebrates, and more recently in murine studies, that classical genetic complementation experiments are often difficult to interpret and are prone to overinterpretation when longevity is the phenotypic readout. Given that longevity is an extraordinarily complex trait and that the regulators of maximum longevity are still poorly understood, it has proven difficult to demonstrate true complementarity in life span studies. There always exists a possibility that some unrelated factor or process is limiting additive life span extension in experiments that appear to demonstrate complementation. Even considering this caveat, there is a large body of literature arising from multiple model organisms that link mTORC1-regulated processes, including autophagy and mRNA translation, to the beneficial effects of CR; together, these strongly suggest a role for mTOR in the CR response. Given the clear links of CR to processes downstream of mTOR, the observed decrease in mTORC1 signaling on CR, and the role of mTOR in nutrient signaling, there is a general consensus that altered mTOR signaling plays a significant role in CR, though additional pathways undoubtedly contribute to the overall effects of the intervention.

As a major energy and growth signaling sensor, mTORC1 acts as a central coordinator of proliferative and maintenance programs. mTORC1 activity drives growth through activation of mRNA translation, regulation of metabolic pathways including glycolysis and fatty acid metabolism, and repression of cellular catabolic pathways, primarily the autophagy/lysosomal degradation pathway. Inhibition of mTORC1 results in reduced mRNA translation, increased catabolic processes, and a shift in metabolic substrate preference. Many of the effects of mTORC1 activity or inhibition are mediated by activation or suppression of downstream transcriptional regulators and the complex crosstalk between these factors. This is particularly apparent in the effects of mTORC1 modulation on metabolism, with the outcomes being highly context dependent on organism, tissue type, duration of intervention, type and severity of intervention, and complex interactions with extracellular or extraorganismal environment. Thus, dissecting out the pathways and targets of key importance in aging and disease is a significant challenge. This is a context in which future work using systems biology approaches may play an especially important role.

While mTOR signaling is complex, some highly conserved and well-defined pathways have been identified downstream of mTORC1 and mTORC2. mTORC1 has been studied more extensively, and mTORC1-regulated processes are generally better described. The best-described mTOR-driven processes are briefly addressed below.

mRNA Translation

Hormonal signaling and abundant nutrient availability promote mTORC1 activation which upregulate a variety of cellular processes necessary for growth. One critical process driven by active mTORC1 signaling is mRNA translation, required for protein synthesis and cell growth. mTORC1 kinase activity is known to promote translation through at least two distinct substrates [15]. mTORC1 phosphorylates p70S6K, the 70-kDa ribosomal S6 kinase, which is an activator of ribosome biogenesis. The interaction of this level of regulation of translation with the control of ribosomal biogenesis by TORC2 has been previously mentioned. Eukaryotic translation initiation factor 4E-binding protein 1 (eIF4E-BP1 or 4E-BP1) is also directly phosphorylated by mTORC1. Phosphorylation of eIF4E-BP1 results in its release from the eukaryotic translation initiation factor 4E (eIF4E), allowing eIF4E to associate with mRNA cap binding proteins and form the cap-dependent translation initiation complex. The formation of this cap-binding complex is a key translation initiation event in eukaryotes.

Activation of mRNA translation is a critical function of mTORC1 and likely accounts for many of the phenotypes associated with mTOR-driven disease, while decreased translation likely mediates many of the positive effects of mTOR inhibition. The antiproliferative effects of mTOR inhibitors in cancer and immune diseases may largely be attributed to decreased rates of protein synthesis due to reduced translation. Decreased mRNA translation has been identified as a major pro-longevity intervention in multiple organisms, and CR is thought to largely act through decreased translation. Deletion or knockdown of ribosomal components or translation initiation factors has been clearly demonstrated to increase life span in yeast, flies and nematodes. Furthermore, deletion of S6K extends life span and decreases body size in mice [6], though rates of translation have not yet been directly examined in these animals.

While mRNA translation is globally decreased in the setting of CR, and in at least some reports of mTOR inhibition, multiple studies in model systems suggest that the beneficial effects of reduced translation may result from differential translation of a subset of mRNAs rather than simply being a consequence of reduced global translation. This model has been best established in budding yeast, where life span extension resulting from ribosomal protein subunit deletion has been linked to an increase in translation of the transcription factor Gcn4 [16]. Gcn4 regulates a variety of genes including genes coding for proteins necessary for response to low nutrient conditions and genes encoding proteins involved in stress response. In yeast, Gcn4 has been found to be necessary for life span extension by ribosomal mutants, and appears necessary for full life span extension by mTOR or S6K deletion. Similar observations have been described in nematodes and flies, but this model has not yet found support in mammalian systems. A recent report using the mTORC1-specific catalytic inhibitor Torin 1 in p53$^{-/-}$ mouse embryonic fibroblasts suggested that differential effects on translation resulting from mTOR inhibition in mammalian cell culture could be largely explained by the presence of a 5′ terminal oligopyrimidine motif, though the study could not rule out the existence of less abundant differentially regulated 5′ or 3′ elements [17]. The interpretation of this study is complicated by the authors' claim that all of the effects of mTOR inhibition were mediated through the 4E-BPs. Deletion of individual 4E-BPs has not been reported to alter translation and has no obvious effect on body size in mice, although it does appear to alter adipose tissue mass [18], while deletion of S6K results in a marked reduction in body mass, demonstrating that S6K regulates growth rate (presumably through direct effects on translation) in vivo. These apparent discrepancies may be a result of the cell culture system or mode of mTOR inhibition – it may be that catalytic inhibition of mTORC1 by Torin 1 does not accurately model mTOR inhibition by rapamycins or through genetic modulation.

Global reductions in mRNA translation may directly contribute to the beneficial effects of mTOR inhibition on age-related diseases involving proteotoxic stress. The reduction of translation rates may directly enhance the fidelity of translation [19], and it is widely accepted that decreases in protein synthesis rates may allow for improved cellular proteostasis through a decreased workload on endogenous protein repair and degradation. A decreased steady state requirement for protein repair and degradation machinery may result in an increased capacity for cells to respond to transient stresses such as oxidative damage, protein aggregation, and heat or cold shock. Loss of proteostasis is a critical component of a number of age-related diseases (see 'mTOR and Age-Related Diseases') and maintaining proteostasis is crucial for organism survival.

It seems that decreased mRNA translation may be promoting longevity at least partially through improved proteostasis and increased protein degradation, though it is difficult to dissect this phenotype away from elevations in autophagy (see below), antioxidant defense, or other biological effects of mTOR inhibition.

Autophagy

In addition to promoting protein synthesis and cell growth, active mTOR inhibits the intracellular catabolic process of autophagy. As a major intracellular recycling pathway in eukaryotes, the autophagy-lysosomal pathway plays an essential role in degrading damaged organelles and macromolecules. Nutrient deprivation decreases mTOR activity, relieving the inhibition of autophagy by active mTOR and resulting in an increase in the catabolism of proteins and organelles. This increased catabolic activity provides amino acids and allows for cell survival when nutrients are limiting.

The observed accumulation of damaged and aggregated proteins, oxidized lipids, and damaged organelles with age suggest that basal levels of autophagy decline or are insufficient to prevent the accumulation of damaged macromolecules associated with aging. Lipofuscin, the complex granular pigment that accumulates in aged tissue, is a highly conserved phenotype of cellular aging that has been observed in all multicellular eukaryotes [20]. While the exact composition and functional consequences of lipofuscin remain to be determined, it has become clear that longevity-promoting interventions also slow the rate of lipofuscin accumulation. Thus, lipofuscin is often used as a biomarker of relative age. Given the close correlation between longevity and damaged macromolecule accumulation, the obvious prediction is that accumulated damaged macromolecules are a driving factor in aging and modulation of this accumulation could attenuate aging. While this hypothesis has proven difficult to test directly given the challenges in selectively inducing the autophagy-lysosomal system, a large body of evidence from yeast and *C. elegans* supports the model that induction of autophagy is a necessary downstream effector of mTORC1 inhibition in mediating life span extension [21]. Induction of autophagy has also been shown to be necessary for CR-mediated longevity promotion, potentially through mTOR [22]. While the necessity of autophagy for the success of these interventions is broadly accepted, it is not clear whether induction of autophagy alone is sufficient to increase life span.

In addition to the role of autophagy in promoting longevity, dysfunction of this pathway has been implicated in a variety of pathologies, and activation of autophagy has been demonstrated to attenuate a variety of age-related diseases. The induction of autophagy has been directly implicated as a potential clinical target for treatment of cardiovascular disease, age-related macular degeneration, diabetes, and a variety of neurodegenerative disorders including Parkinson's disease (PD) and Alzheimer's disease (AD) [23]. The nervous system appears particularly sensitive to the accumulation of damaged macromolecules and protein aggregates, and increasing autophagic degradation has been shown to prevent neurodegeneration in models of AD and PD as well as in models of Huntington's disease (HD), a progressive neurodegenerative disease directly associated with proteotoxic insult.

Mitochondrial Function and Metabolism

Mitochondria are key organelles in metabolism, disease, and aging. These organelles are the major producers of energy for most cell types, a primary site of metabolic reactions, a major source of toxic products (both reactive oxygen species and toxic intermediates of metabolism) and provide key cellular signaling regulators. Given the multifaceted role that mitochondria play in eukaryotic biology, it is unsurprising that mitochondria have been linked to a variety of pathological states, diseases, and aging [24]. mTORC1 appears to influence mitochondrial function through multiple mechanisms and downstream regulatory factors.

Hypoxia-inducible factor 1, Hif-1, is a transcription factor that promotes glycolytic processes, and can be activated through mTOR signaling in mammals [25]. This factor is linked to longevity in model organisms through somewhat unclear mechanisms and to vascular tumor growth, 'wet' macular degeneration [26], and rheumatoid arthritis [27] in mammals through its positive effects on the angiogenic factor VEGF. At the intracellular level, Hif-1 promotes glycolysis, downregulates mitochondrial oxygen consumption, and at least partially mediates the Warburg effect in mammalian neoplasia. Decreased mTOR signaling is thought, therefore, to directly influence tumor vascularization, tumor metabolism, and the Warburg effect at least partially through decreased activation of Hif-1. This pathway may act independently of, or cooperatively with the general anti-proliferative effects of mTOR inhibition.

Consistent with these effects, mTOR inhibition has been associated with increased mitochondrial respiration in yeast and worms, and CR has been associated with increased mitochondrial content and respiration in a variety of organisms. This effect has been directly associated with longevity in yeast, with adaptive signaling resulting from increased mitochondrial superoxide production being implicated in this response [28]. Mice lacking mTORC1 components in white adipose tissue also show an increase in mitochondrial content and respiration, suggesting that this may be conserved in mammals [29]. Mitochondrial metabolism and cellular mitochondrial mass have also been reported to be increased by mTORC1 inhibition, at least in certain conditions, through a downstream activation of PGC-1α and the transcription factor Ying-Yang 1 [30]. The precise role of mitochondrial metabolism as a downstream mediator of mTOR is far from clear, but available data suggest that this is a critical component of aging and disease, and thus warrants further attention.

In addition to the above, mTOR affects mitochondrial function through autophagic degradation of mitochondria, a process termed mitophagy. The mitochondrial-lysosomal axis theory of aging suggests that proper maintenance of a functional pool of mitochondria depends on continuous successful removal of damaged and dysfunctional mitochondrial components through fission and mitophagy of fission products

[31]. This theory predicts that reducing the rate of turnover of mitochondria, as may occur in aging, would result in an accumulation of dysfunctional mitochondria. This accumulation could lead to an increase in basal ROS production, damage accumulation, loss of tissue homeostasis, and potentially cellular senescence or death. mTORC1 inhibition increases basal autophagy, as described above, and would thus be predicted to preserve or restore mitochondrial function with age and thus potentially improve overall mitochondrial function. While this model remains to be directly addressed, it provides an attractive link between mitochondrial function and mTOR signaling in aging and disease.

Stem Cell Maintenance

Stem cell loss and dysfunction are likely a significant factor in mammalian aging and age-related diseases. This is particularly likely in proliferative tissues such as dermis, the immune and gastrointestinal system, as well as in wound repair or response to ischemic injury in which proliferative capacity is required. While the exact role for stem cells in aging is currently unknown, evidence suggests that mTORC1 is central in the maintenance of stem cells with age. As discussed below, mTOR inhibition has been shown to protect immune function with age in murine models of infection, and this has been attributed to enhanced hematopoietic stem cell capacity. mTOR inhibition with rapamycin has also been recently reported to improve intestinal stem cell function, although in this case the improvement was linked to alterations in mTOR signaling in the adjacent Paneth cells, which are responsible for maintaining the stem cell niche, rather than a direct effect on intestinal stem cells. CR has been shown to enhance the function of skeletal muscle stem cells, presumably related, at least in part, to the concomitant decrease in mTOR activity. The role of stem cells in aging and of mTOR in regulating their function remains an exciting and largely uncharted avenue of research.

mTOR and Age-Related Disease

Longevity and Health Span

Extension of health span, defined as the duration of life for which an organism is free from major age-related disease or loss of function, is considered by many to be the critical goal of aging research. Longevity studies in model organisms can intrinsically include health span components. *C. elegans* viability determination in standard plate or liquid-based life span studies is based on the ability of the animals to respond to mechanical stimulus. Yeast studies, both replicative and chronological, depend on the cell capacity to successfully produce progeny. In both cases, the individual organism may remain viable beyond the point that they are considered deceased by the assay standards. Thus, the nematode and yeast models are tied to neurological, muscular health, and reproductive capacity, respectively. In the murine model animal welfare regulations generally prevent expiration of mice by natural causes, requiring eutha-nasia if animals decline past a set cutoff in body mass, appear immobile, hunched, or in pain, or if they show signs of severe and untreatable diseases, such as ulcers or can-cers. These restrictions may complicate accurate determination of life span in a lon-gevity study but they also compel longevity-promoting interventions to be those that, at least in part, also protect health span.

Considering these restrictions, it is perhaps unsurprising that current data suggest longevity-enhancing interventions extend the health of populations and decrease or delay the incidence of age-related disease, rather than increase survival of unhealthy individuals. It appears that the regulation of health span and life span are at least closely linked. Concordantly, relatively few examples exist in the literature where health span is benefited in the absence of longevity benefits, perhaps a result of using life span as a primary end point. A noted exception is the recent (2013) NIA study of CR in rhesus monkeys, which observed a significant decrease in the appearance of age-related diseases without a change in survival [32]. While this, and studies like it, show that it may be possible to uncouple health span and longevity, they are gener-ally limited by a lack of positive controls and/or a clear definition of what constitutes baseline health span. The rhesus study, for example, stands in contrast to a prior study that reported an increase in both life span and health span [33]. Differences in diet, housing conditions, and severity of the CR protocol may be factors distinguishing the two studies and the lack of a positive control limits the conclusions that can be drawn from this work. It is possible that longevity-promoting interventions may more sen-sitively and broadly affect health span than life span.

There has been a recent growth in emphasis on efforts to define and characterize the effects of longevity-promoting interventions on age-related health parameters. Each model used for aging studies has a set of health parameters that have been used to ex-plore the relationship between life span, health span, and aging interventions. In yeast, the primary health span parameters are replicative capacity (in this case the actual read-out for life span), mitochondrial function, and cell morphology. Nematodes have typi-cally been used to study proteostasis and clearance of damaged macromolecules during the aging process as well as neurological decline, diseases of proteotoxicity, and age-related changes in muscle function. *D. melanogaster* has been useful in studying neuro-logical function, muscle function, sensory function, stem cell function, and cardiac function with age. Mammalian systems have been used to examine a variety of age-re-lated physiological and health parameters relevant to the biology of human aging.

mTOR and Disease

Concurrent with demonstrations of a role for mTOR in regulating longevity, it be-came increasingly apparent that mTOR signaling plays a central role in regulating health span and a variety of age-related and non-age-related pathologies (table 1).

Table 1. mTOR inhibition in age-related diseases

Disease	Observation
Neurodegenerative	
PD	Protection of dopaminergic neurons in fly and mouse models
AD	Delayed disease progression in mouse models
HD	Induction of autophagy and reduction in htt toxicity in fly and mouse models
Age-related cognitive decline	Improved spatial learning and memory in old mice; antidepressive effects
Cancer	Inhibition of growth of solid tumor cell lines; disappointing efficacy in clinical trials with the exception of renal cell carcinoma and several rare cancers that include those driven by TSC1 and TSC2 deficiency
Heart disease	
Restenosis following angioplasty	Widespread use of stents that elute rapamycins
Hypertrophy and failure	Inhibition of hypertrophy and some regression of failure in the mouse aortic constriction model
Cardiomyopathy	Zebrafish heterozygous for mTOR were protected in two different models of cardiomyopathy; rapamycin attenuated cardiomyopathy in LMNA$^{-/-}$ mice
Diabetes and obesity	
Protection	Mice lacking S6k1 in all tissues or lacking Raptor in adipose tissue are resistant to diet-induced obesity and insulin insensitivity
Susceptibility	Mice and rats chronically treated with rapamycin develop glucose intolerance and insulin resistance, attributed to inhibition of mTORC2; associated with rapamycin in transplantation therapy in humans
Immune function	
Inhibition	Use in clinical immunosuppression therapies
Enhancement	Enhanced immune function in tuberculosis and anti-tumor vaccine responses in mice and vaccinia vaccination in nonhuman primates Improved B cell and influenza vaccine responses in old mice
Kidney disease	
Allografts and renal cancer	Reduced rejection and nephrotoxicity with rapamycin therapy
Polycystic kidney disease	Improved outcome in animal models
Age-related macular degeneration	Decreased the incidence and severity of retinopathy in rats; decreased need for anti-VEGF treatment in humans
Hutchinson-Gilford progeria	Corrected the nuclear morphology defect, delayed onset of cellular senescence, and enhanced clearance of progerin through autophagic degradation in cell culture models

Altered mTOR signaling has been investigated as a potential therapeutic strategy in a range of age-related and age-associated pathologies.

mTOR inhibition attenuates specific age-related changes in lower organisms, as noted above. In addition, mTOR inhibition slows or delays many age-related and age-associated changes that are broadly conserved from lower eukaryotes to mammals. Included among these are lipofuscin accumulation, DNA damage accumulation, age-related mitochondrial dysfunction, and loss of proteostasis. Cardiac, neuronal, and stem cell functions are also improved with age in multicellular invertebrates and mammals by inhibition of mTOR. While the efficacy of mTOR inhibitors in attenuating age-related pathologies in humans is yet to be determined, there is an abundance of literature suggesting that mTOR is a clear potential target for diseases of aging.

Heart Disease

There is evidence that mTOR inhibition may be generally protective against cardiomyopathies, including age-related cardiomyopathy [34]. Zebrafish heterozygous for mTOR were protected against two different models of cardiomyopathy. Administration of rapamycin markedly suppresses cardiac hypertrophy in the trans-aortic constriction model of pressure overload-induced heart failure, and rapamycin treatment has been shown to result in regression of established pressure overload-induced cardiac hypertrophy, fibrosis, and dysfunction. Perhaps the greatest impact of mTOR-targeted pharmacotherapy in cardiac disease has emerged through the widespread use of stents that elute rapamycin or rapamycin derivatives (e.g. everolimus, temsirolimus, ridaforolimus, umirolimus, zotarolimus, collectively referred to as 'rapamycins') to inhibit cell proliferation and restenosis following angioplasty, with significant decreases in major adverse cardiovascular events during the first few years after implantation [35].

While decreased mTOR signaling has been clearly demonstrated to attenuate a variety of cardiac myopathy and failure models, the precise mechanisms of importance are less clear [34]. Decreased mRNA translation, inflammation, and hypertrophic growth signaling, increased autophagy, improved mitochondrial function or content, and altered metabolic preference have all been independently linked to improved outcome resulting from mTOR inhibition in cardiac models making it difficult to parse out the key functions downstream of mTOR crucial to the benefits observed. The complex nature of the cardiac system and the multifaceted role of mTOR inhibition in attenuating cardiac dysfunction make age-related cardiac dysfunction a particularly attractive model for a systems approach. Much of the recent and ongoing work in the cardiac aging field relies on systems biology techniques, such as shotgun proteomics, RNA sequencing, and metabolomics, to demonstrate a shift in organ state in aging and an attenuation with treatment.

These stand-alone systems analyses are proving highly informative for understanding the effects of interventions, but a full mechanistic picture of the role of mTOR in age-related cardiac disease and rescue will likely require a combination of multiple approaches. Systems analyses will likely reveal multiple distinct functional states in cardiac tissue where interventions affecting any of the individual downstream mediators of mTOR actually lead to a similar shift in the steady state through feedback. Decreased translation and increased autophagy could, in this scenario, converge in their functional consequence by causing an overall shift in the functional, proteomic or metabolic state of cardiac cells. Thus, the cardiac aging paradigm seems particularly amenable to a systems biology perspective.

Neurodegenerative Disease

As noted above, it has been suggested that mTORC1 inhibition-mediated enhancement of autophagy may lead to improved degradation of aberrant or misfolded proteins and reduced proteotoxic stress in neurodegenerative diseases such as PD, AD, and HD [36]. Thus, inhibition of mTOR could prove to be a successful therapeutic strategy in neurodegenerative disease. Evidence of such effects has been seen in fly and murine models of PD as well as in fly, murine, and cell culture models of HD. Positive effects of rapamycin on disease progression have been reported in two different mouse models of AD.

The evidence of a benefit of rapamycin in neurodegenerative diseases raises the question of whether mTORC1 inhibition might also attenuate age-related declines in cognitive function in the absence of a more severe neurological disorder. Recent studies assessing the effects of chronic mTOR inhibition on cognitive function during aging in mice have reported that old animals treated with rapamycin performed substantially better on tasks measuring spatial learning and memory than did untreated, age-matched animals [37]. Intriguingly, there were indications that rapamycin also enhanced cognitive function in young mice and had anti-anxiety and antidepressive effects at all ages. mTOR has also been identified as a potential target for treatment of seizures [38], suggesting that growth signaling inhibition may have broad neurological benefits. The complexity of the central nervous system makes it a challenge to investigate the mechanisms underlying the beneficial effects of decreased mTOR signaling, but the clear therapeutic potential make it a very attractive target for testing mTOR inhibitors in a clinical setting.

Cancer

A majority of tumors show evidence for activation or upregulation of mTOR signaling, and mTOR inhibition has been studied extensively as a potential therapy for a wide variety of cancers. Rapamycins potently inhibit growth of solid tumor cell lines but have shown disappointing efficacy in several clinical trials, though certain rare cancers, including renal cell carcinomas and glioblastomas, may respond to mTOR inhibition. Mutations in TSC1 and TSC2, which are upstream inhibitors of mTOR, cause a variety of hyperplastic diseases including tuberous sclerosis, directly linking hyperplasia to mTOR overactivity. Dysplasias driven by TSC1 or TSC2 loss are currently being evaluated as clinical targets for mTOR inhibitors [39]. Rapamycin has been shown to delay or reduce deaths due to age-related and age-associated cancers in several studies in mice.

Diabetes and Obesity

mTOR signaling has been implicated in the development of age-associated metabolic disorders such as obesity and type 2 diabetes. Inhibition of mTORC1 inhibits, while mTORC1 activation stimulates, adipogenesis in mice [40]. Obesity results in chronic activation of mTOR in adipose tissue, a state that has been linked to obesity-associated cancers, inflammation, β-cell adaptation preceding type 2 diabetes, nonalcoholic fatty liver disease, and many other complications [41]. Multiple downstream effectors of mTOR signaling, such as S6K-1, 4EBP1, and SREBP, act as mediators between nutrient signaling and the development of obesity and type 2 diabetes. SREBP, a transcription factor that induces the expression of lipogenic genes, is activated by mTORC1 but not mTORC2 [41]. S6K-1 null mice display reduced body fat mass and resistance to diet-induced obesity, while mice lacking 4EBP1 show increased sensitivity to diet-induced obesity and adipogenesis, possibly through hyperactivation of S6K-1 [41]. Mice lacking Raptor in adipose tissue are lean with fewer and smaller adipocytes, have increased insulin sensitivity, and show resistance to diet-induced obesity [29]. Conversely, mice with adipose-specific knock out of Rictor have normal body fat mass and glucose tolerance but are hyperinsulinemic [42]. Mice lacking S6K1 in all tissues or lacking RAPTOR specifically in adipose tissue show a profound resistance to diet-induced obesity [29].

The relationship between mTOR signaling and age-related metabolic disorders, including type 2 diabetes and obesity, is complicated. Rapamycin can protect mice from diet-induced obesity through the inhibition of adipocyte differentiation [43]. However, mice and rats chronically treated with rapamycin demonstrate altered metabolic homeostasis, observed as altered insulin sensitivity and glucose tolerance [44]. These effects of chronic rapamycin have recently been attributed to effects on mTORC2. Blood hyperlipidemia is also observed with chronic rapamycin treatment in mouse and human studies [40]. The S6K1 knockout mice are also hypoinsulinemic and glucose intolerant, apparently due to a decrease in β-cell size and function. Despite these diabetes-like symptoms, both S6K1 knockout mice and rapamycin treated mice are long-lived, suggesting that these effects are not so detrimental as to limit survival. It has also been pointed out that the 'starvation-induced diabetes' associated with mTORC1 inhibition differs substantially from type 2 diabetes, which is caused by insulin resistance resulting from overnutrition in association with mTOR activation. Thus, additional study is needed to determine whether targeted inhibition of mTORC1 and the observed changes in lipid profile, insulin sensitivity, and glucose homeostasis represent a health risk or an altered metabolic state consistent with the promotion of longevity and health span.

Immune Function
Rapamycins are used clinically as immunosuppressive or immunomodulatory drugs. There are, however, also reports that rapamycins can enhance immune system efficacy in certain settings, including tuberculosis, anti-tumor vaccine responses in mice, and vaccinia vaccination in nonhuman primates [45]. In the context of age-related immune function, treating 22- to 24-month-old mice with rapamycin for only 6 weeks doubled the percentage and number of B (but not T) cells in the bone marrow, and restored the capacity of the aged animals' immune system to mount an effective response to influenza vaccination, which was protective against subsequent

infection. The apparent contradiction between these observations and the use of rapamycins as immunosuppressive drugs may be explained by observations that mTOR can exert divergent immunoregulatory functions during immune cell activation and differentiation, depending on the cell subset type. Furthermore, while rapamycins may limit immune activation and proliferation in the setting of an immunogenic insult, they appear to have a robust effect in preventing age-related declines in immune function, thus preserving immune function later in life. Thus, rapamycins' functions in immune biology are more complex than previously recognized, with outcomes depending on dose, duration of treatment, immune cell type, and specific immune challenge. Furthermore, the long-term effects of rapamycins on age-related declines in immune function stand in stark contrast to those of short-term responses.

Inflammation

Inflammation is strongly associated with aging and drives a multitude of age-associated disorders. Cardiovascular disease, obesity and metabolic disorders, cancer, and neurodegenerative diseases all include inflammatory components, and attenuation of inflammation has been implicated as a clinical target in each of these disease settings. Hyperactive mTOR has been linked to inflammation, and inhibition of mTOR by rapamycins has been demonstrated to be anti-inflammatory in renal disease, lung infection, and in vascular inflammation in atherosclerosis and following angioplasty. Furthermore, CR strongly attenuates age-related inflammatory signaling, and this effect appears to be at least partly mediated through mTOR. Reduced mTOR thus seems a good candidate for treating or preventing age-related inflammatory processes, while reduced inflammation seems to play significant mechanistic role in the pro-longevity effects of mTOR inhibitors.

Renal Disease

Rapamycins are used clinically to reduce nephrotoxicity in chemotherapy, prevent allograft rejection, and as a treatment for renal cell carcinoma. Activation of mTOR signaling has been associated with several common forms of kidney disease, suggesting that inhibition of mTOR might have broad therapeutic benefits for renal health. Consistent with this, rapamycins have been shown to reduce kidney fibrosis, attenuate diabetic nephropathy, and improve outcome in animal models of polycystic kidney disease.

Age-Related Macular Degeneration

Age-related macular degeneration is the leading cause of blindness in Western countries. Capillary overgrowth in the choroid layer of the eye, which is a contributing factor, has been attributed to excessive production of VEGF. Rapamycin has been shown to reduce VEGF expression in retinal pigment epithelium and inhibit angiogenesis in vitro [46]. In a rat model of age-related macular degeneration, rapamycin decreased the incidence and severity of retinopathy [47] and in human patients rapamycin appeared to decrease the need for anti-VEGF intravitreal injections by approximately half [48]. Thus, age-related macular degeneration appears to be a promising clinical target for mTOR-inhibiting interventions.

Hutchinson-Gilford Progeria and Laminopathies
Hutchinson-Gilford progeria syndrome (HGPS) is typically caused by a de novo mutation in the lamin A/C gene (LMNA) that activates a cryptic splice site, producing an abnormal lamin A protein termed progerin. Accumulation of progerin leads to aberrant nuclear morphology in vitro, and is believed to be the causal factor in the pathogenesis of disease. The precise mechanism linking progerin accumulation to the phenotypes associated with this disease is unclear, but it is generally thought to involve disruption of nuclear DNA binding proteins, including transcription factors and DNA repair components, as a result of aberrant nuclear scaffold structure. It has been reported that treatment of cells from HGPS patients with rapamycin corrects the nuclear morphology defect, delays the onset of cellular senescence, and enhances the clearance of progerin through autophagic degradation [49]. No effective treatment for HGPS currently exists, and these data provide hope that rapamycins might slow disease progression in HGPS patients. Any success in HGPS would strongly suggest that rapamycins might show efficacy in patients diagnosed with atypical Werners' syndrome, often caused by non-HGPS mutations in LMNA, as well as in patients with muscular dystrophies resulting from lamin mutations or other laminopathies.

Systems Biology Approaches to Studying mTOR

Genome-wide approaches, such as yeast single-gene mutant screens and RNAi screens in nematodes, have been critical to uncovering genes involved in the regulation of life span. The advent of proteomics, metabolomics, whole-genome sequencing, and RNAseq has fundamentally altered the way that aging studies are designed and executed. As the accessibility of these high-throughput methods has improved, these techniques are being increasingly utilized. The study of mTOR in aging has historically been largely led by reductionist experiments, with researchers focusing primarily on the use of single gene or small molecule perturbations to study mTOR and aging in model organisms. While these approaches have been, and continue to be, very fruitful in defining the regulation of aging and age-related processes by mTOR, they are limited in their ability to model the complexities of aging.

As large datasets produced using these methods become more widely available, it becomes increasingly important that the systems biology paradigm is applied to produce a comprehensive picture of the information (fig. 2). Modern aging research will greatly benefit from collaborations with bioinformatics experts that extend be-

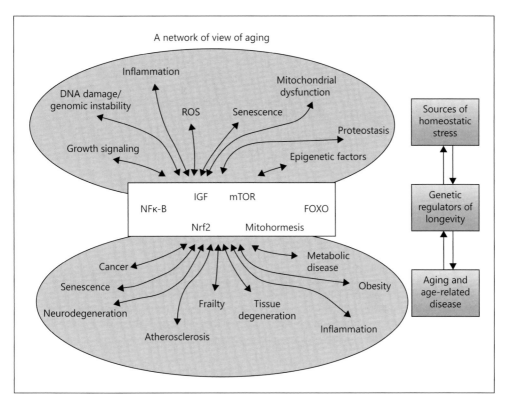

Fig. 2. A network view of aging. A network approach to aging emphasizes the extensive crosstalk among and between the underlying causes of age-related changes, age-related diseases and dysfunction, and the cellular mechanisms and mediators that regulate longevity. This paradigm highlights the importance of systems biology approaches as a perturbation to any individual component in these networks is likely to affect many other components, and the overall outcome depends on the interactions between all players involved.

yond analysis of high-throughput data and into systems-based integration of diverse datasets into models that combine genetic, proteomic, transcriptomic, and metabolomics data into informative and approachable descriptions of aging processes. These models should provide novel targets and strategies for intervention in the aging process as new nodes are described in genetic, proteomics, transcriptional, and metabolic paradigms. In addition, the systems approach to aging research may provide clear answers to difficult scientific queries, such as the nature of the similarities and differences between CR and CR mimetics (such as rapamycin). Thus, while classic methods for studying aging are far from exhausted, it is clear that systems approaches have a role to play in our understanding of aging, the influence of genotype on aging, and the mechanisms of interventions, such as mTOR inhibition, on the aging process.

References

1 Cafferkey R, Young PR, McLaughlin MM, Bergsma DJ, Koltin Y, Sathe GM, Faucette L, Eng WK, Johnson RK, Livi GP: Dominant missense mutations in a novel yeast protein related to mammalian phosphatidylinositol 3-kinase and VPS34 abrogate rapamycin cytotoxicity. Mol Cell Biol 1993;13:6012–6023.

2 Chiu MI, Katz H, Berlin V: RAPT1, a mammalian homolog of yeast TOR, interacts with the FKBP12/rapamycin complex. Proc Natl Acad Sci USA 1994; 91:12574–12578.

3 Fabrizio P, Pozza F, Pletcher SD, Gendron CM, Longo VD: Regulation of longevity and stress resistance by Sch9 in yeast. Science 2001;292:288–290.

4 Jia K, Chen D, Riddle DL: The TOR pathway interacts with the insulin signaling pathway to regulate *C. elegans* larval development, metabolism and life span. Development 2004;131:3897–3906.

5 Johnson SC, Rabinovitch PS, Kaeberlein M: mTOR is a key modulator of ageing and age-related disease. Nature 2013;493:338–345.

6 Kenyon CJ: The genetics of ageing. Nature 2010;464: 504–512.

7 Huang J, Manning BD: The TSC1-TSC2 complex: a molecular switchboard controlling cell growth. Biochem J 2008;412:179–190.

8 Xie X, Guan KL: The ribosome and TORC2: collaborators for cell growth. Cell 2011;144:640–642.

9 Gwinn DM, Shackelford DB, Egan DF, Mihaylova MM, Mery A, Vasquez DS, Turk BE, Shaw RJ: AMPK phosphorylation of raptor mediates a metabolic checkpoint. Mol Cell 2008;30:214–226.

10 Björntorp P, Carlström S, Fagerberg SE, Hermann LS, Holm AG, Scherstén B, Ostman J: Influence of phenformin and metformin on exercise-induced lactataemia in patients with diabetes mellitus. Diabetologia 1978;15:95–98.

11 Jewell JL, Russell RC, Guan KL: Amino acid signalling upstream of mTOR. Nat Rev Mol Cell Biol 2013; 14:133–139.

12 Mercken EM, Carboneau BA, Krzysik-Walker SM, de Cabo R: Of mice and men: the benefits of caloric restriction, exercise, and mimetics. Ageing Res Rev 2012;11:390–398.

13 Swindell WR: Dietary restriction in rats and mice: a meta-analysis and review of the evidence for genotype-dependent effects on lifespan. Ageing Res Rev 2012;11:254–270.

14 Hansen M, Taubert S, Crawford D, Libina N, Lee SJ, Kenyon C: Lifespan extension by conditions that inhibit translation in *Caenorhabditis elegans*. Aging cell 2007;6:95–110.

15 Proud CG: mTORC1 signalling and mRNA translation. Biochem Soc Trans 2009;37:227–231.

16 Steffen KK, MacKay VL, Kerr EO, Tsuchiya M, Hu D, Fox LA, Dang N, Johnston ED, Oakes JA, Tchao BN, Pak DN, Fields S, Kennedy BK, Kaeberlein M: Yeast life span extension by depletion of 60s ribosomal subunits is mediated by Gcn4. Cell 2008;133: 292–302.

17 Thoreen CC, Chantranupong L, Keys HR, Wang T, Gray NS, Sabatini DM: A unifying model for mTORC1-mediated regulation of mRNA translation. Nature 2012;485:109–113.

18 Tsukiyama-Kohara K, Poulin F, Kohara M, DeMaria CT, Cheng A, Wu Z, Gingras AC, Katsume A, Elchebly M, Spiegelman BM, Harper ME, Tremblay ML, Sonenberg N: Adipose tissue reduction in mice lacking the translational inhibitor 4E-BP1. Nature 2001; 7:1128–1132.

19 Conn CS, Qian SB: Nutrient signaling in protein homeostasis: an increase in quantity at the expense of quality. Sci Signal 2013;6:ra24.

20 Cho S, Hwang ES: Fluorescence-based detection and quantification of features of cellular senescence. Methods Cell Biol 2011;103:149–188.

21 Rubinsztein DC, Mariño G, Kroemer G: Autophagy and aging. Cell 2011;146:682–695.

22 Jia K, Levine B: Autophagy and longevity: lessons from *C. elegans*. Adv Exp Med Biol 2010;694:47–60.

23 Nixon RA, Yang DS: Autophagy and neuronal cell death in neurological disorders. Cold Spring Harb Perspect Biol 2012;4:pii a008839.

24 Lee HC, Wei YH: Mitochondria and aging. Adv Exp Med Biol 2012;942:311–327.

25 Wouters BG, Koritzinsky M: Hypoxia signalling through mTOR and the unfolded protein response in cancer. Nat Rev Cancer 2008;8:851–864.

26 Grimm C, Willmann G: Hypoxia in the eye: a two-sided coin. High Alt Med Biol 2012;13:169–175.

27 Konisti S, Kiriakidis S, Paleolog EM: Hypoxia – a key regulator of angiogenesis and inflammation in rheumatoid arthritis. Nat Rev Rheumatol 2012;8:153–162.

28 Ristow M, Zarse K: How increased oxidative stress promotes longevity and metabolic health: the concept of mitochondrial hormesis (mitohormesis). Exp Gerontol 2010;45:410–418.

29 Polak P, Cybulski N, Feige JN, Auwerx J, Rüegg MA, Hall MN: Adipose-specific knockout of raptor results in lean mice with enhanced mitochondrial respiration. Cell Metab 2008;8:399–410.

30 Cunningham JT, Rodgers JT, Arlow DH, Vazquez F, Mootha VK, Puigserver P: mTOR controls mitochondrial oxidative function through a YY1-PGC-1alpha transcriptional complex. Nature 2007;450: 736–740.

31 Terman A, Kurz T, Navratil M, Arriaga EA, Brunk UT: Mitochondrial turnover and aging of long-lived postmitotic cells: the mitochondrial-lysosomal axis theory of aging. Antioxid Redox Signal 2010;12:503–535.

32 Mattison JA, Roth GS, Beasley TM, Tilmont EM, Handy AM, Herbert RL, Longo DL, Allison DB, Young JE, Bryant M, Barnard D, Ward WF, Qi W, Ingram DK, de Cabo R: Impact of caloric restriction on health and survival in rhesus monkeys from the NIA study. Nature 2012;489:318–321.

33 Colman RJ, Anderson RM, Johnson SC, Kastman EK, Kosmatka KJ, Beasley TM, Allison DB, Cruzen C, Simmons HA, Kemnitz JW, Weindruch R: Caloric restriction delays disease onset and mortality in rhesus monkeys. Science 2009;325:201–204.

34 Dai DF, Chen T, Johnson SC, Szeto H, Rabinovitch PS: Cardiac aging: from molecular mechanisms to significance in human health and disease. Antioxid Redox Signal 2012;16:1492–1526.

35 Kim U, Park JS, Lee SH, Shin DG, Kim YJ: Seven-year clinical outcomes of sirolimus-eluting stent versus bare-metal stent: a matched analysis from a real world, single center registry. J Korean Med Sci 2013;28:396–401.

36 Chong ZZ, Shang YC, Wang S, Maiese K: Shedding new light on neurodegenerative diseases through the mammalian target of rapamycin. Prog Neurobiol 2012;99:128–148.

37 Halloran J, Hussong SA, Burbank R, Podlutskaya N, Fischer KE, Sloane LB, Austad SN, Strong R, Richardson A, Hart MJ, Galvan V: Chronic inhibition of mammalian target of rapamycin by rapamycin modulates cognitive and non-cognitive components of behavior throughout lifespan in mice. Neuroscience 2012;223:102–113.

38 Galanopoulou AS, Gorter JA, Cepeda C: Finding a better drug for epilepsy: the mTOR pathway as an antiepileptogenic target. Epilepsia 2012;53:1119–1130.

39 Ruiz-Falcó Rojas ML: Therapeutic update in tuberous sclerosis complex: the role of mTOR pathway inhibitors. Rev Neurol 2012;54(suppl 3):S19–S24.

40 Laplante M, Sabatini DM: mTOR signaling in growth control and disease. Cell 2012;149:274–293.

41 Laplante M, Sabatini DM: An emerging role of mTOR in lipid biosynthesis. Curr Biol 2009;19:R1046–R1052.

42 Cybulski N, Polak P, Auwerx J, Rüegg MA, Hall MN: mTOR complex 2 in adipose tissue negatively controls whole-body growth. Proc Natl Acad Sci USA 2009;106:9902–9907.

43 Bell A, Grunder L, Sorisky A: Rapamycin inhibits human adipocyte differentiation in primary culture. Obes Res 2009;8:249–254.

44 Lamming DW, Ye L, Katajisto P, Goncalves MD, Saitoh M, Stevens DM, Davis JG, Salmon AB, Richardson A, Ahima RS, Guertin DA, Sabatini DM, Baur JA: Rapamycin-induced insulin resistance is mediated by mTORC2 loss and uncoupled from longevity. Science 2012;335:1638–1643.

45 Turner AP, Shaffer VO, Araki K, Martens C, Turner PL, Gangappa S, Ford ML, Ahmed R, Kirk AD, Larsen CP: Sirolimus enhances the magnitude and quality of viral-specific CD8+ T-cell responses to vaccinia virus vaccination in rhesus macaques. Am J Transplant 2011;11:613–618.

46 Stahl A, Paschek L, Martin G, Gross NJ, Feltgen N, Hansen LL, Agostini HT: Rapamycin reduces VEGF expression in retinal pigment epithelium (RPE) and inhibits RPE-induced sprouting angiogenesis in vitro. FEBS Lett 2008;582:3097–3102.

47 Kolosova NG, Muraleva NA, Zhdankina AA, Stefanova NA, Fursova AZ, Blagosklonny MV: Prevention of age-related macular degeneration-like retinopathy by rapamycin in rats. Am J Pathol 2012;181:472–477.

48 Nussenblatt RB, Byrnes G, Sen HN, Yeh S, Faia L, Meyerle C, Wroblewski K, Li Z, Liu B, Chew E, Sherry PR, Friedman P, Gill F, Ferris F 3rd: A randomized pilot study of systemic immunosuppression in the treatment of age-related macular degeneration with choroidal neovascularization. Retina 2010;30:1579–1587.

49 Graziotto JJ, Cao K, Collins FS, Krainc D: Rapamycin activates autophagy in Hutchinson-Gilford progeria syndrome: implications for normal aging and age-dependent neurodegenerative disorders. Autophagy 2012;8:147–151.

Peter S. Rabinovitch, MD, PhD
Department of Pathology, University of Washington, Box 357705
1959 NE Pacific St., K-081 HSB
Seattle, WA 98195 (USA)
E-Mail PeterR@medicine.washington.edu

Yashin AI, Jazwinski SM (eds): Aging and Health – A Systems Biology Perspective.
Interdiscipl Top Gerontol. Basel, Karger, 2015, vol 40, pp 128–140 (DOI: 10.1159/000364975)

Melatonin and Circadian Oscillators in Aging – A Dynamic Approach to the Multiply Connected Players

Rüdiger Hardeland

Johann Friedrich Blumenbach Institute of Zoology and Anthropology, University of Göttingen, Göttingen, Germany

Abstract

From the perspective of systems biology, melatonin is relevant to aging in multiple ways. As a highly pleiotropic agent, it acts as a modulator and protectant of mitochondrial electron flux, a potent antioxidant that supports the redox balance and prevents excessive free radical formation, a coregulator of metabolic sensing and antagonist of insulin resistance, an immune modulator, a physiological hypnotic and, importantly, an orchestrating chronobiotic. It entrains central and peripheral circadian clocks and is required for some high-amplitude rhythms. The circadian system, which controls countless functions, is composed of many cellular oscillators that involve various accessory clock proteins, some of which are modulated by melatonin, e.g. sirtuin 1, AMP-dependent protein kinase, and protein kinase Cα. Aging and age-related diseases are associated with losses in melatonin secretion and rhythm amplitudes. The dynamic properties of aging processes deserve particular attention. This concerns especially two vicious cycles, one of peroxynitrite formation driven by inflammation or overexcitation, another one of inflammaging driven by the senescence-associated secretory phenotype, and additionally the loss of dynamics in a deteriorating circadian multioscillator system.

© 2015 S. Karger AG, Basel

Among scientific terms, 'homeostasis' is one of those which are very frequently used in an inappropriate manner. Of course, homeostasis reflects a concept of considerable value. Plasma levels of insulin and of Ca^{2+} are classic examples of its usefulness. However, even basically homeostatically regulated parameters can be modulated by a minor but demonstrable circadian component, as known since long for the cases of both insulin [1, 2] and Ca^{2+} [3, 4]. From a cybernetic point of view, homeostasis is based on a negative feedback loop that allows the system to approach a setpoint value. However, with a sufficient delay time in the system, the feedback may result in an oscillation. Therefore, the existence of a negative feedback loop does not at all imply that this would lead to a near-constant level. Oscillations may also be superimposed from outside the feedback circuit. An impressive example is the level of plasma cortisol [5], which exhibits a circadian rhythm with one of the highest amplitudes among blood parameters, despite a feedback to the upper instances in the hypothalamo-hypophyseal-adrenal axis. Contrary to the earlier belief that the delay time of the feedback may contribute to rhythmicity and that corresponding upstream rhythms in CRH and ACTH are causal to the glucocorticoid rhythm, the latter is meanwhile known to be generated by a peripheral circadian oscillator present in the adrenal cortex [6]. At least in mice, this oscillator requires the presence of melatonin for generating robust rhythms [7].

The idea that the functioning of an organism is mainly based on the principle of homeostasis, i.e. a dynamic equilibrium attained by feedback mechanisms, turns out to be a misconception. A closer look reveals that countless processes in the body are rhythmic. The 'circadian organization' is apparent in functions as different as behavior, mood, sleep/wakefulness, other neuronal activities, muscular strength, hormones, cytokines, intraorganismal distribution and actions of various immune cells, circulation, vegetative functions, intracellular structure, chromatin remodeling, protein expression, enzyme activities, and many parameters more.

Our growing insight into the significance of circadian oscillators and their output functions for health and healthy aging is currently increasingly perceived [8, 9]. This includes the notion that genes of the cellular core oscillators act as tumor suppressors, that tumors suppress their intracellular circadian oscillators by epigenetic means, that an aging suppressor such as sirtuin 1 (SIRT1) and other metabolic sensors turn out to be accessory oscillator proteins that interact with the core oscillator, and that mutations in core oscillator genes lead to increased formation of free radicals and to cancer.

This perception is insofar of particular importance as the circadian oscillator system, which is composed of numerous central and peripheral oscillators with a different degree of autonomy and sensitivity to external time cues, controls a high number of physiological and cell biological processes. A specific aspect concerns melatonin, which is mainly secreted by the pineal gland, but is also synthesized in numerous other tissues and cells [10]. The fraction released from the pineal gland exhibits a high-amplitude circadian rhythm with a prominent nocturnal peak. The actions of this indoleamine display an unusually high degree of pleiotropy [10]. The melatonin receptors MT_1 and/or MT_2 are expressed in many cell types, not only in those with high receptor density such as the hypothalamic suprachiasmatic nucleus (SCN), the central circadian master clock. In mammals, a dual relationship between SCN and melatonin exists insofar as melatonin secretion is steered by the SCN, and melatonin feeds back to the SCN, thereby influencing the circadian phase and, in diurnally active mammals such as the human, initiating sleep [11]. In addition, melatonin effects have been demonstrated in several peripheral circadian oscillators, and are assumed in others [8].

In relation to aging, both the SCN and melatonin levels exhibit changes presumed to be unfavorable and to accelerate processes of senescence [9, 12]. Circadian amplitudes often decrease by age, which may lead, e.g., to sleep disturbances and nocturia. Typically, the spontaneous circadian period changes during aging, with the consequence of phase advances under synchronized conditions. In the extreme, rhythms may decompose into separate, differently coupled components. In many but not all individuals, nocturnal melatonin levels decrease considerably, and the rhythm may almost disappear at advanced age. These changes are even more pronounced in a number of diseases and disorders, especially in neurodegenerative pathologies [12]. With regard to the multiple connections of the circadian system and of melatonin to numerous functions, these alterations are of particular gerontological interest. Under the aspect of cyclicity, the significance of these changes can be only understood from a dynamical perspective.

A Systemic View on the Interconnections of Major Aging Processes

From the perspective of systems biology, it is important to perceive the multiple connections between aging processes. A selection of the most important causes of damage and alterations resulting hereof are depicted in figure 1. In particular, the various sources of reactive oxygen and nitrogen (RNS) species, their interaction in terms of peroxynitrite formation and the consequences for mitochondrial function, cell proliferation and the immune system are emphasized in this overview. Of course, areas of high complexity such as immunosenescence can be touched in this scheme only superficially. Age-dependent changes in the immune system including immune remodeling, inflammaging and the new insights concerning the senescence-associated secretory phenotype (SASP) have been recently discussed in relation to melatonin and circadian rhythmicity [9].

Vicious Cycles

In addition to the circadian rhythmicity, which is involved in manifold ways in free radical formation, detoxification and avoidance, in mitochondrial metabolism and in the immune system, dynamic processes are also initiated and amplified by positive feedback loops, which can promote deteriorating changes during senescence. In figure 2a, the crucial role of peroxynitrite ($OONO^-$) and free radicals formed from this compound is outlined. As soon as $\cdot NO$, the free-radical congener of nitric oxide, is formed at higher rates, especially as a consequence of neuronal overexcitation or inflammation, a simultaneously occurring elevated generation of superoxide ($O_2\cdot^-$) unavoidably leads to peroxynitrite because superoxide has a similar affinity to $\cdot NO$ and to the superoxide dismutases, the detoxificants of $O_2\cdot^-$. Protonation of peroxynitrite leads to the unstable acid, which readily decomposes to the highly reactive, devastating hydroxyl radical ($\cdot OH$) and $\cdot NO_2$. An additional reaction exists, which is frequent-

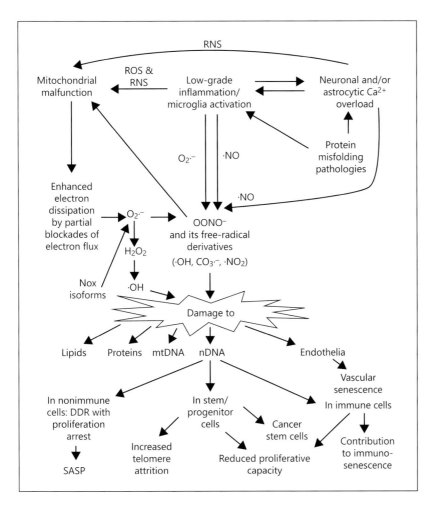

Fig. 1. Simplified overview of some aging processes. DDR = DNA damage response; mtDNA = mitochondrial DNA; nDNA = nuclear DNA; Nox = NAD(P)H oxidase; ROS = reactive oxygen species.

ly underrated in its importance. Peroxynitrite forms an adduct with CO_2, which decomposes in a corresponding way to a carbonate radical ($CO_3\cdot^-$) and $\cdot NO_2$. Although the carbonate radical has a lower reactivity than $\cdot OH$, it undergoes similar oxidative reactions and has, by virtue of resonance stabilization, a longer lifetime than the extremely rapidly decaying $\cdot OH$ and is, thus, farther reaching. In conjunction with either $\cdot OH$ or $CO_3\cdot^-$, $\cdot NO_2$ leads to nitration of aromates, including tyrosine residues in proteins [13]. Further reactions are described in this reference. It seems important to be aware of the role of CO_2 in this context since it is highly available in mitochondria where it is formed as well as in tissues and in the circulation under conditions of hypoperfusion, a cause of enhanced tyrosine nitration, endothelial damage and vascular senescence. Notably, melatonin is an efficient scavenger of carbonate radicals [14] and a mitochondrial protectant [9, 13].

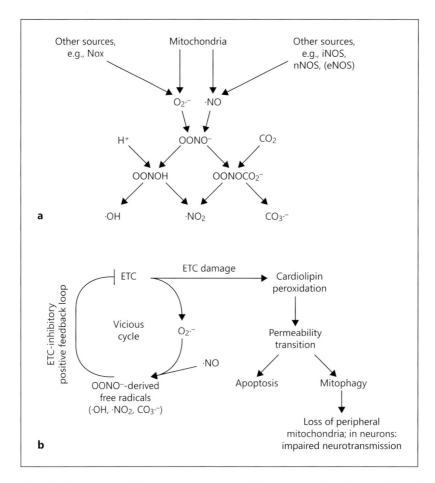

Fig. 2. Role of peroxynitrite and a vicious cycle driving progressive damage of the electron transport chain (ETC), apoptosis and mitophagy. **a** Formation of free radicals from peroxynitrite. **b** Vicious cycle. The feedback loop that inhibits electron fluxes, in cybernetic terms, a positive one because it enhances electron leakage. iNOS = Inducible NO synthase; nNOS = neuronal NO synthase; eNOS = endothelial NO synthase.

The formation of peroxynitrite is part of a vicious cycle with relevance to aging (fig. 2b). RNS are known to interrupt electron flux at different points of the mitochondrial electron transport chain (ETC). In the extreme of high-grade inflammation, it can completely block the entire pathway. Under conditions of low-grade inflammation, the damage by peroxynitrite-derived free radicals is crucial to the impairment of electron flux. The resulting bottlenecks cause an enhanced rate of electron leakage via electron back- and overflow mainly at complexes I and III [for further details, see Hardeland 15]. Dissipating electrons are transferred to molecular oxygen to give superoxide. As long as ·NO formation takes place at an enhanced rate, the increased formation of superoxide and, therefore, peroxynitrite causes a steady amplification of ETC dysfunction and radical generation through this vicious cycle. ETC damage ul-

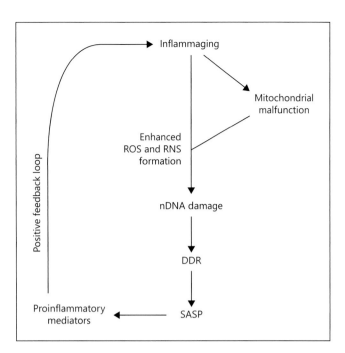

Fig. 3. The vicious cycle of inflammaging with SASP.

timately leads to cardiolipin peroxidation by the peroxidase activity of the cytochrome c/cardiolipin complex [16], permeability transition and apoptosis. Interestingly, melatonin acts in nontumor cells as an anti-apoptotic agent and a regulator of permeability transition that allows short-term opening of the transition pore, but inhibits its persistent opening [for details, including discussion of superoxide flashes and differences to tumor cells, see Hardeland 9]. Alternately, mitochondrial damage can lead to mitophagy. On the one hand, this rescues cells from death, but, on the other hand, it can cause losses especially in the numbers of peripheral mitochondria, what is frequently observed during aging and impairs neurotransmission at synapses insufficiently supplied with ATP.

A second vicious cycle is also related to free radicals and inflammation, but involves damage to the nuclear DNA (fig. 3). This initiates the DNA damage response, which includes a proliferative arrest, a mechanism that prevents a carcinogenic fate of the affected cells. However, these senescent cells display the previously unexpected property of secreting, even as nonimmune cells, numerous factors including proinflammatory cytokines. This so-called SASP represents a driving force of aging-related inflammation. Even at low grade, this mechanism causes additional oxidative and nitrosative damage and can become an undesired source of carcinogenesis, although the primary action of arresting cells represents an anticarcinogenic action [17, 18]. In conjunction with shifts from anti-inflammatory to proinflammatory cytokine secretion that occur as a consequence of immune remodeling, SASP largely contributes to inflammaging.

It is important to be aware that vicious cycles behave in a highly dynamic way, far from homeostatic system properties. Because mitochondrial metabolism, components of the antioxidative protection system including melatonin, neuronal activities and many aspects of the immune system undergo circadian changes [19, 20], the vicious cycles are intertwined with a second dynamic system driven by cellular oscillators.

Beneficial Cycles

Circadian rhythms are not only a means for structuring our day/night-related activities and for anticipating physiological requirements to come a few hours ahead, but are truly beneficial with regard to health and, thus, healthy aging. Some of these aspects had not been foreseen in the past. The important finding that mice carrying mutations in the core oscillator gene *Per2* are cancer prone [21] has been later extended to other core oscillator genes, which have been identified as tumor suppressor genes [8]. This role is, among other effects, related to circadian cycles of chromatin remodeling, which include histone acetylation by the CLOCK protein and deacetylation by SIRT1, by upregulation of other tumor suppressor genes such as *Wee-1*, and by suppression of protooncogenes such as *c-myc* [summarized in Hardeland et al. 8]. Mutations in genes of the core oscillator and associated factors that cause deviations in period length or make the oscillator dysfunctional have been shown to increase the damage by free radicals, in organisms as different as hamsters and *Drosophila*, to enhance the susceptibility to exogenously induced oxidative stress, and may be related to the observation that repeated experimental phase shifts reduce the lifetime of *Drosophila* [details in Hardeland et al. 8, 19]. These findings are believed to reflect the rhythmicities of both free radical formation and detoxification.

The cyclicity of the highly complex circadian system has manifold implications for the optimal functioning of a body, for health and aging. In figure 4, an overview is presented for some major roles of circadian oscillators. It combines findings obtained in different central and peripheral oscillators. The oscillators of different cells are not entirely identical, although they are operating on the basis of the same principle. Even in a single tissue, different clocks exist, which are acting in parallel and utilize different orthologs or even paralogs of the core oscillator proteins. Moreover, the various accessory oscillator proteins are sometimes cell type specific. Finally, input pathways for synchronizing time cues as well as the degree of autonomy can differ. Some of them are more sensitive to the timing of food intake, whereas others strongly depend on the light/dark cycle. In figure 4, a selection of physiological or cell biological functions is summarized that are relevant to health and aging. Some of the output functions are feeding back to the respective oscillators and are capable of reentraining circadian rhythms. Among these, melatonin plays a particularly important role, because it modulates or synchronizes a plethora of circadian output functions including

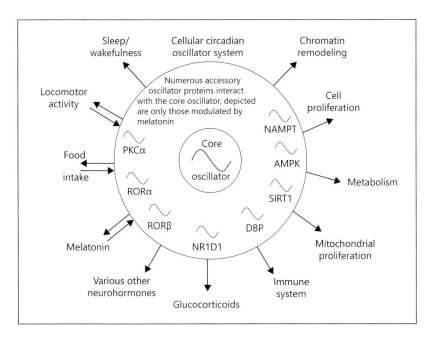

Fig. 4. Schematic representation of several accessory oscillator proteins that can be modulated by melatonin in cellular circadian clocks and a selection of output functions, some of which feed back to the oscillator. The various central and peripheral oscillators differ with regard to the presence of depicted details. AMPK = AMP-dependent protein kinase; DBT = D site of albumin promoter binding protein; NAMPT = nicotinamide phosphoribosyltransferase; NR1D1 = nuclear receptor subfamily 1, group D, member 1; PKCα = protein kinase Cα; RORα, RORβ = retinoic acid receptor-related orphan receptor-α, -β.

those depicted in figure 4. A remarkable aspect concerns the numerous accessory oscillator proteins, among which all those have been incorporated that are also under control by melatonin.

Pleiotropy of Melatonin in the Context of Aging

After the discovery of the antioxidant and cell protective properties of melatonin [22], the questions arose as to why a nocturnally peaking agent should be effective in diurnally active mammals such as the human, which generate more free radicals during the day, and whether results obtained in nocturnal laboratory rodents can be applicable to man. Part of the answer is that direct scavenging of free radicals requires high melatonin concentrations for being efficacious and is only relevant where sufficient levels are attained, e.g. in melatonin-forming cells, perhaps in melatonin-accumulating organelles such as mitochondria [9, 10], and at pharmacological levels. The other part of the answer is related to the coordination and phasing of circadian rhythms, presumably also to a support of high-amplitude oscillations as observed, at least, in

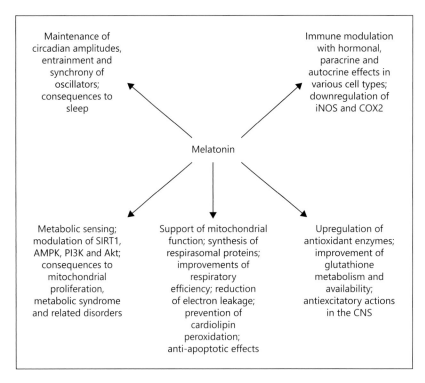

Fig. 5. Actions exerted by the highly pleiotropic regulator molecule melatonin, a selection with particular relevance to aging. CNS = Central nervous system; COX2 = cyclooxygenase 2; PI3K = phosphatidylinositol 3-kinase.

several cases. In conjunction with other effects, such as prevention of neuronal over-excitation and reduction of mitochondrial electron leakage, the concept of radical avoidance [23] was proposed, assuming a higher significance of the prevention of enhanced radical generation compared to the detoxification of radicals already formed.

Melatonin exerts a number of effects of importance to aging, as summarized in figure 5. The support and coordination of rhythms is not only of relevance to the avoidance of excessive damage by free radicals, as observed under conditions of disturbed rhythmicity in oscillator gene mutants. A sufficiently high and appropriately timed increase in melatonin is also implicated in sleep initiation [12]. This involves an action at the SCN with a downstream effect on the hypothalamic sleep switch and an additional thalamic action that initiates, via a thalamocortical interplay, the formation of sleep spindles. Sleep disturbances represent a highly frequent change and complaint associated with aging and can have a number of secondary consequences concerning, e.g., nutrition, insulin resistance, and changes in the immune system.

The actions of melatonin are also intertwined with the pathways of metabolic sensing (fig. 5). Effects on AMP-dependent protein kinase, phosphatidylinositol 3-kinase and Akt have been repeatedly described under various conditions, as recently summarized [9]. Upregulation of SIRT1, an aging suppressor, metabolic sensor and accessory oscillator protein, have been reported in a few cases related to aging, but the reverse was found in tumor cells [cf. 9]. Effects on mitochondrial proliferation may be associated with metabolic sensors and downstream factors such as peroxisome proliferator-activated receptor-γ coactivator-1α and peroxisome proliferator-activated receptor-γ [9]. Beneficial effects by melatonin and synthetic melatonergic agonists on metabolic syndrome, insulin resistance and diabetes type 2 have been repeatedly described and reviewed [8, 9, 12]. Effects of melatonin on mitochondria exceed the aspect of their proliferation and intracellular distribution, as partially addressed in a preceding section. Further details have been elaborated in normally aging and senescence-accelerated animals, as summarized elsewhere [9, 12]. These findings obtained in mitochondria include the upregulation of antioxidant enzymes, improved formation and availability of reduced glutathione, and ·NO metabolism.

Finally, melatonin formed by the pineal gland and by leukocytes is related to immune functions in multiple ways. Actions of melatonin in the diverse subtypes of immune cells and concerning the secretion of numerous cytokines have been described [9, 10] and are relevant to immune remodeling during aging and the particularly important aspect of inflammaging [9]. The problem of both anti- and proinflammatory actions exerted by melatonin is addressed in the next section.

Controversial Results at First but Not at Second Glance: The Importance of Circadian Dynamics

Among the numerous actions of melatonin, several reports on seemingly opposite effects appear, at first glance, controversial or, as soon as they have been repeatedly confirmed, at least paradoxical. However, they may turn out to be less implausible when regarded from the perspective of systems biology and, where appropriate, under consideration of dynamic changes as well as interventions that block dynamic processes.

One example for opposite effects concerns melatonin's immunological actions. Both anti-inflammatory and proinflammatory effects of the methoxyindole have been described. In the majority of immune cells, melatonin has stimulatory properties [10]. Therefore, it is not surprising that proinflammatory and, thus, pro-oxidant effects have also been described. However, what matters is the balance between pro- and anti-inflammatory actions and the conditions under which the balance is shifted. As a rule, anti-inflammatory actions are most evident under conditions of high-grade inflammation, especially endotoxemia and sepsis. The protective property of melatonin is largely related to suppression of excessive ·NO production, reduction of peroxynitrite formation, maintenance of glutathione levels and improvement of mitochondrial function. Proinflammatory effects are frequently observed under basal conditions. Notably, melatonin does not suppress basal or moderately enhanced ·NO formation by inducible NO synthase or neuronal NO synthase and, therefore, does not prevent these stimulatory actions. The immunological role of melatonin may, thus, be understood as that of a buffering agent [24], which allows moderate upregulations but sets limits to excessive, damaging processes [9]. However, this interpretation may not fully explain what is found under conditions of senescence. Especially with regard to concerns related to the promotion of inflammatory processes in autoimmune diseases, which become more frequent in the course of immunosenescence, one might have expected a primarily detrimental role of melatonin as an immune stimulator during aging. However, the opposite is observed, in both normally aging and senescence-accelerated rodents, in which proinflammatory cytokines were downregulated and anti-inflammatory mediators upregulated [summarized in Hardeland 9]. In addition, reductions of oxidative damage and improvements of mitochondrial activity and efficiency were reported. Although the reasons for the predominantly anti-inflammatory actions may require further clarification, the interruption of vicious cycles (cf. fig. 2, 3) may be of particular importance. Improvements of mitochondrial function and antioxidative actions of melatonin may prevent an enhanced release of inflammatory mediators and, thus, normalize the system.

Another area in which opposite effects of melatonin have been observed concerns apoptosis. Numerous publications have demonstrated anti-apoptotic effects of melatonin in nontumor cells, in vivo and in culture. These comprise increases in anti-apoptotic proteins such as Bcl-2 or Bcl-x_L, decreases in their proapoptotic counterparts such as Bad and Bax, inhibition of Bad dephosphorylation and of poly-ADP ribose polymerase cleavage, prevention of cardiolipin peroxidation, of permanent mitochondrial permeability transition, cytochrome c release and caspase-3 activation [summarized in Hardeland 9, and Hardeland et al. 10]. However, several more recent reports have shown that melatonin can also behave in a proapoptotic way [9, 25]. However, these surprising findings have been made, at reasonable melatonin concentrations, in tumor cells. This difference has been even observed by comparing a pseudo-tumorigenic tissue, the human primary villous trophoblast, in which melatonin remained antiapoptotic, with choriocarcinoma cells [25]. Similarly contrasting findings, with particular importance to aging, were obtained when effects of melatonin on SIRT1 expression were studied [9]. Again, the sirtuin was found to be upregulated in normal cells, especially also in comparisons of young and aged neurons, and in senescence-accelerated mice, but it was reported to be downregulated in tumor cells [26; for further literature see Hardeland 9]. Again, this melatonin effect in cancer cells was associated with the induction of apoptosis. The causes of these differences between tumor and nontumor cells require further clarification. However, it seems important to consider in this context their differences in circadian dynamics. As mentioned in the introductory section, tumors can suppress their cellular circadian oscillators epigenetically, by promoter hypermethylation in core oscillator genes, and thereby block the tumor suppressor function of oscillator proteins. Therefore, it would be of high relevance to know in which phase state the oscillator is trapped. This could imply that certain clock proteins are maintained at an expression minimum and others at their maximum. Since SIRT1 acts as an accessory oscillator protein, it may be fixed at a certain relatively high expression level when the oscillator is stopped in a tumor cell. An agent like melatonin that interacts with circadian clocks may push the system into another phase position, e.g. by the known inhibition of *Rorα* expression and subsequent downregulation of the core oscillator protein BMAL1, even if the oscillator is in total not yet sufficiently operating. In other words, effects of melatonin may be entirely different depending on the circadian phase and can be completely divergent in operating and arrested oscillators. This interpretation may lead to other, more general implications of relevance to experimental approaches. Overexpression of clock proteins may likewise lead to unphysiological states of oscillators that may not allow unambiguous conclusions.

Conclusion

For the understanding of aging, the consideration of dynamic processes is required. This includes the circadian oscillator system and the melatonin rhythm in their aging- and disease-related deterioration and, moreover, vicious cycles of peroxynitrite formation driven by inflammation or overexcitation and of inflammaging driven by the SASP. Melatonin has the potential of readjusting rhythms and breaking the vicious cycles.

References

1 Lambert AE, Hoet JJ: Diurnal pattern of plasma insulin concentration in the human. Diabetologia 1966;2:69–72.

2 Nelson W, Bingham C, Haus E, Lakatua DJ, Kawasaki T, Halberg F: Rhythm-adjusted age effects in a concomitant study of twelve hormones in blood plasma of women. J Gerontol 1980;35:512–519.

3 Parfitt AM: Bone and plasma calcium homeostasis. Bone 1987;8(suppl 1):S1–S8.

4 Fijorek K, Puskulluoglu M, Polak S: Circadian models of serum potassium, sodium, and calcium concentrations in healthy individuals and their application to cardiac electrophysiology simulations at individual level. Comput Math Models Med 2013; 2013:429037.

5 Goldman J, Waichenberg BL, Liberman B, Nery M, Achando S, Germek OA: Contrast analysis for the evaluation of the circadian rhythms of plasma cortisol, androstenedione, and testosterone in normal men and the possible influence of meals. J Clin Endocrinol Metab 1985;60:164–167.

6 Chung S, Son GH, Kim K: Adrenal peripheral oscillator in generating the circadian glucocorticoid rhythm. Ann NY Acad Sci 2011;1220:71–81.

7 Torres-Farfan C, Serón-Ferré M, Dinet V, Korf HW: Immunocytochemical demonstration of day/night changes of clock gene protein levels in the murine adrenal gland: differences between melatonin-proficient (C3H) and melatonin-deficient (C57BL) mice. J Pineal Res 2006;40:64–70.

8 Hardeland R, Madrid JA, Tan D-X, Reiter RJ: Melatonin, the circadian multioscillator system and health: the need for detailed analyses of peripheral melatonin signaling. J Pineal Res 2012;52:139–166.

9 Hardeland R: Melatonin and the theories of aging: a critical appraisal of melatonin's role in antiaging mechanisms. J Pineal Res 2013;55:325–356.

10 Hardeland R, Cardinali DP, Srinivasan V, Spence DW, Brown GM, Pandi-Perumal SR: Melatonin – a pleiotropic, orchestrating regulator molecule. Prog Neurobiol 2011;93:350–384.

11 Stehle JH, von Gall C, Korf HW: Melatonin: a clock-output, a clock-input. J Neuroendocrinol 2003;15: 383–389.

12 Hardeland R: Melatonin in aging and disease – multiple consequences of reduced secretion, options and limits of treatment. Aging Dis 2012;3:194–225.

13 Hardeland R: Melatonin and its metabolites as anti-nitrosating and anti-nitrating agents. J Exp Integ Med 2011;1:67–81.

14 Hardeland R, Poeggeler B, Niebergall R, Zelosko V: Oxidation of melatonin by carbonate radicals and chemiluminescence emitted during pyrrole ring cleavage. J Pineal Res 2003;34:17–25.

15 Hardeland R: Melatonin, mitochondrial electron flux and leakage: recent findings and resolution of contradictory results. Adv Stud Biol 2009;1:207–230.

16 Kagan VE, Bayir HA, Belikova NA, Kapralov O, Tyurina YY, Tyurin VA, Jiang J, Stoyanovsky DA, Wipf P, Kochanek PM, Greenberger JS, Pitt B, Shvedova AA, Borisenko G: Cytochrome c/cardiolipin relations in mitochondria: a kiss of death. Free Radic Biol Med 2009;46:1439–1453.

17 Fumagalli M, d'Adda di Fagagna F: SASPense and DDRama in cancer and ageing. Nat Cell Biol 2009; 11:921–923.

18 Coppé JP, Desprez PY, Krtolica A, Campisi J: The senescence-associated secretory phenotype: the dark side of tumor suppression. Annu Rev Pathol 2010;5: 99–118.

19 Hardeland R, Coto-Montes A, Poeggeler B: Circadian rhythms, oxidative stress and antioxidative defense mechanisms. Chronobiol Int 2003;20:921–962.

20 Scheiermann C, Kunisaki Y, Frenette PS: Circadian control of the immune system. Nat Rev Immunol 2013;13:190–198.

21 Fu L, Pelicano H, Liu J, Huang P, Lee C: The circadian gene Period2 plays an important role in tumor suppression and DNA damage response in vivo. Cell 2002;111:41–50.

22 Reiter RJ: Oxidative damage in the central nervous system: protection by melatonin. Prog Neurobiol 1998;56:359–384.

23 Hardeland R: Antioxidative protection by melatonin – multiplicity of mechanisms from radical detoxification to radical avoidance. Endocrine 2005;27: 119–130.

24 Carrillo-Vico A, Lardone PJ, Álvarez-Sánchez N, Rodríguez-Rodríguez A, Guerrero JM: Melatonin: buffering the immune system. Int J Mol Sci 2013;14: 8638–8683.

25 Lanoix D, Lacasse AA, Reiter RJ, Vaillancourt C: Melatonin: the smart killer: the human trophoblast as a model. Mol Cell Endocrinol 2012;348:1–11.

26 Jung-Hynes B, Schmit TL, Reagan-Shaw SR, Siddiqui IA, Mukhtar H, Ahmad N: Melatonin, a novel Sirt1 inhibitor, imparts proliferative effects against prostate cancer cells in vitro culture and in vivo in TRAMP model. J Pineal Res 2011;50:140–149.

Rüdiger Hardeland
Johann Friedrich Blumenbach Institute of Zoology and Anthropology
University of Göttingen, Berliner Strasse 28
DE–37073 Göttingen (Germany)
E-Mail rhardel@gwdg.de

Yashin AI, Jazwinski SM (eds): Aging and Health – A Systems Biology Perspective.
Interdiscipl Top Gerontol. Basel, Karger, 2015, vol 40, pp 141–154 (DOI: 10.1159/000364976)

Diet-Microbiota-Health Interactions in Older Subjects: Implications for Healthy Aging

Denise B. Lynch[a, b] · Ian B. Jeffery[a, b] · Siobhan Cusack[a] ·
Eibhlis M. O'Connor[c] · Paul W. O'Toole[a, b]

[a]School of Microbiology and [b]Alimentary Pharmabiotic Centre, University College Cork, Cork, and
[c]Department of Life Sciences, University of Limerick, Limerick, Ireland

Abstract

With modern medicine and an awareness of healthy lifestyle practices, people are living longer and generally healthier lives than their ancestors. These successes of modern medicine have resulted in an increasing proportion of elderly in society. Research groups around the world have investigated the contribution of gut microbial communities to human health and well-being. It was established that the microbiota composition of the human gut is modulated by lifestyle factors, especially diet. The microbiota composition and function, acting in concert with direct and indirect effects of habitual diet, is of great importance in remaining healthy and active. This is not a new concept, but until now the scale of the potential microbiota contribution was not appreciated. There are an estimated ten times more bacteria in an individual than human cells. The bacterial population is relatively stable in adults, but the age-related changes that occur later in life can have a negative impact on host health. This loss of the adult-associated microbiota correlates with measures of markers of inflammation, frailty, co-morbidity and nutritional status. This effect may be greater than that of diet or in some cases genetics alone. Collectively, the recent studies show the importance of the microbiota and associated metabolites in healthy aging and the importance of diet in its modulation.

© 2015 S. Karger AG, Basel

The human gastrointestinal tract has multiple critical roles that are central to health and well-being. It serves as a way to process foods and eliminate waste. It helps control appetite, it houses the largest mucosal surface in the body housing a large component of the immune system, and it even contributes to mood, behaviour and general well-being. The most numerate cellular part of the gastrointestinal tract is not human but microbial. Most of these bacteria are called commensals, meaning they are normally resident and metabolizing human dietary components. The microbes in and on the human body have a combined number of genes a hundred times that of the human gene complement. The gastrointestinal tract is one of the most diverse environments on the planet. The intestinal microbiota is now recognized as a major environmental modifier of health risk. Independent of genetic and other lifestyle factors, the gut microbiota has a coding capacity and potential metabolic activity that has a major impact on human physiology. In infancy, the microbiota composition trends towards an adult pattern over the first 2–3 years, with low initial diversity increasing over this time period. Disruptions of this process may be associated with risk for allergic disease in later life. In the adult years, alterations in the microbiota are associated with a diverse range of diseases [reviewed in de Vos and de Vos 1]. There is a particularly compelling case for studying the microbiota in aging subjects. This phase of life is accompanied by a range of physiological and lifestyle changes that can have a big effect on the physical environment of the intestine. It has been known for several decades that the gut microbiota of older persons, similar to the very young, is in a state of flux [2]. Coupled with a wide range of reported alterations in the composition of the intestinal microbiota in seniors, and different rates of age-related health loss in different individuals, countries and populations, detailed analysis of gut microbiota-health interactions in older people is particularly appropriate. This review summarizes the differences between the physiology of older subjects and young adults that are relevant for microbiota changes and details the major findings of culture-based studies, and then examines the health implications of recent culture-independent studies, including those from the largest study to date, the ELDERMET consortium.

Physiological and Clinical Issues That Can Impact on the Gut Microbiota in Elderly

The global proportion of older people is rapidly and continually increasing. This has resulted in an increased need for healthcare and societal supports for this cohort of our society, and has highlighted the importance of not just longevity but healthy aging that maximizes functional capacity and quality of life in older age. The diversity of the microbiota of an individual is shaped by a number of factors, both internal to the host, and external. Common, age-related, physiological changes can modify physiological function, which can in turn alter the composition of the microbiota.

Physiological, motor and sensory functions change with age. For instance, a natural reduction in dentition and deteriorating muscle mass in later life can impact on mastication ability. This can limit dietary choices, and changes in the diet can greatly impact the microbiota. Aging may be accompanied by impairment of intestinal sensation and consequently increased susceptibility to gastrointestinal complications. Other age-related, digestive system complications include dysphagia (difficulty swallowing), functional dyspepsia (painful, difficult or disturbed digestion), gastroesophageal reflux, delayed intestinal transit time, diverticulosis, and increased rates of constipation, faecal and gaseous incontinence, all of which can significantly impact on microbiota composition and host health. Importantly, the impairment of taste and thirst sensation, olfaction and digestion, coupled with malabsorption and an increase in the levels of satiation in older people, can lead to imbalances in nutrient intake, malnutrition and significant perturbation of the microbiota [3].

Bilateral interaction with the host facilitates functional conditioning of the immune system by the microbiota, which influences the composition of the microbiota itself. Microbiota disturbance has been linked with an increased susceptibility to disorders including allergies, cancer, digestive/intestinal disorders, frailty, obesity/metabolic disorder and its related conditions. It can also affect regulatory systems such as hormone signalling, leading to changes in mood and behaviour. Host metabolic pathways that facilitate connection between the intestine and the brain, can be affected [3]. Disruption to this bidirectional homeostatic pathway has been associated with inflammation and alterations in the stress response, among other stress-related symptoms such as anxiety, commonly experienced in older age. A healthy, more diverse microbiota composition encourages resistance to pathogens and increased interaction with the host immune system. Loss of diversity in old age is associated with less resistance to pathogens and a natural decline of immune function (immunosenescence) with the development of chronic, low-grade inflammation typical of older age (inflammaging). Both low-diversity microbiota and immunosenescence can lead to increased rates of gastrointestinal infection.

In older age, complex and dynamic exogenous factors, including diet and lifestyle modifications, medication use (particularly antibiotics), disease, injury and stress further influence the composition of the microbiota. Health throughout life, and particularly in later years, is dependent on the maintenance of homeostasis, the presence of a stable physiological environment. The relative stability of the adult intestinal microbiota at a species level is a key contributory factor to the promotion and maintenance of health. However, at abundance level the composition of the microbiota can fluctuate substantially over a short period of time [4]. This suggests that the microbiota is able to respond to exogenous influences throughout life.

Culture-Based Analyses of Intestinal Microbiota of Elderly

Culture-based methods were traditionally used to analyze the intestinal microbiota. An example of some of the methods utilized can be seen from experiments conducted in 1989 in Japan [5]. Culture-based methods were used to compare the microbiota of elderly people in Tokyo, Japan, with elderly in Yuzurihara, an area of Japan where the elderly tend to live longer. The faecal microbiota of 15 healthy elderly subjects from each of the two areas was collected. A number of experiments were performed to determine the genus, and where possible, species, of isolates found in these samples. Serial dilutions in an anaerobic diluent were made, and the samples were subsequently spread onto 4 non-selective and 11 selective agar plates. Subculturing from anaerobic plates to other plates helped determine which microorganisms were strict anaerobes.

In order to identify the isolates, many biochemical tests were performed on broth cultures. These tests include detection of bacteria-derived metabolites and the determination of the effect of bile on bacterial growth. Benno et al. [5] reported that while most of the same genera were observed between the two groups, the Yuzurihara subjects had a larger bifidobacteria contingent than was observed from the Tokyo subjects. However the Yuzurihara subjects had fewer total bacteria, anaerobes, bacilli, clostridia, *Bacteroides* species, and *Eubacterium aerofaciens*. Four genera, *Megamonas*, *Mitsuokella*, *Selenomonas*, and *Acidaminococcus*, were isolated from the Yuzurihara subjects but not the Tokyo subjects. Intestinal bifidobacteria counts are known to decrease with age, and some Enterobacteriaceae increase. That the Yuzurihara elderly had more bifidobacteria than the Tokyo elderly, despite being older, suggests that the Yuzurihara subjects were not displaying the same age-related microbiota changes that we see in other parts of the world. Benno et al. [5] suggest that this is due to the high-fibre diet of the Yuzurihara subjects, and that this is why the Yuzurihara people tend to live longer.

In 2002, another culture-based study focused on elderly suffering from *Clostridium difficile*-associated diarrhoea (CDAD) [6] who had a history of antibiotic treatments resulting in disturbed microbiota. This altered microbiota provided a reduced resistance to *C. difficile* infection. With the widespread use of antibiotics and increasing number of elderly, CDAD has become a challenging problem. Hopkins and Macfarlane [6] aimed to characterize the microbiota of elderly subjects with CDAD. They classified isolates according to their cellular fatty acid profiles. Their results showed that CDAD patients had the lowest species diversity when compared with healthy elderly and young subjects, particularly of bifidobacteria, *Bacteroides* and *Prevotella*. Facultative species were higher in CDAD patients than in healthy subjects. Together, this shows that *C. difficile* is associated with a greatly altered microbiota. The same group completed further studies in 2004 [7], this time comparing healthy young and elderly, and hospitalized elderly. Again, they used fatty acids to identify bacteria. They reported reduced numbers and species diversity in both bifidobacteria and *Bacteroides* in elderly compared with the young subjects.

The benefits of using such culture-based methods for analyzing intestinal microbiota include the low cost, and ability to retain isolates for further analyses. However, there are many disadvantages to culture-based methods. It is labour intensive, and with current approaches it is still not possible to culture the majority of the estimated gut bacteria (estimated 50–90%). Of those species that do grow on current artificial media, certain species will outgrow others, leading to further biases. Another disadvantage of culture-based approaches is difficulties in phylogenetic classifications. For some microbial families, multiple methods must be used to classify genera and species of different families, such as those discussed above. Benno et al. [5] required a large number of methods for classification of isolates. This indicates how complex it can be to identify isolates using culture-based methods. It also shows how much culture of a given isolate is required to identify it. Some of these methods could often not distinguish between two species of a given genus, so biologically-relevant species-specific genes or functions could not be accounted for. Speedy, high-throughput, specific and reliable alternatives were required.

The Technological Revolution

The last decade has seen the introduction of increasingly intense research techniques. Rather than attempting to culture all isolates from a given environment, the DNA can be directly extracted from samples. In theory, this approach provides an unbiased view of the isolates within a sample. The preferred locus used for identification is the 16S ribosomal DNA. As a housekeeping gene found in almost all bacteria, often in high-copy numbers, it is easily amplified. It contains a number of variable regions that differ between species and/or genera and so allows efficient identification.

Real-time quantitative PCR is often used to determine the proportion of certain bacteria in a sample. This approach is fast, cheap and useful for determining the level of a specific group of bacteria. However, when trying to assess and compare a number of different groups, qPCR becomes laborious. This directed approach, while very useful in many cases, does not provide an exhaustive view of the gut population, which is proving to be increasingly important. In 2009, a phylogenetic microarray was developed specifically for the human intestinal tract, known as the HITChip [8]. This chip consists of 4,809 probes, and further probes can be added when required. However, microarrays are a high-throughput targeted approach. While they cover more targets at once, they are still limited by the probes. Different probes have different hybridization abilities, so biases can be introduced based on the choice of probes.

The current, more commonly used technology is high-throughput sequencing. There are a number of different sequencing technologies available; however, microbial community analyses based on 16S ribosomal DNA studies tend to use 454 FLX Titanium pyrosequencing due to the longer reads that can be obtained [9]. Up to 1.6 million reads can be sequenced in one run. Many different samples can be loaded on one slide using barcoded adaptors.

With any new technology such as pyrosequencing, programs for analysis must be developed. The aims are to maximize the data obtained while minimizing the potential for error. Speed and accuracy are paramount, and as increasing amounts of data are obtained, programs that can handle ample quantities of reads are essential. When handling pyrosequencing data, many steps can be executed. Multiplexed libraries must be separated. Adaptor sequences must be removed. Error correction, or denoising, can be performed. Chimeric sequences are sometimes formed during PCR amplification steps. Programs are available to remove these. Clustering is performed to reduce the time and volume for further steps. Finally, sequences must be classified at different phylogenetic levels.

Table 1. Comparing enterotypes with CAGs (co-abundance groups)

Enterotypes	CAGs
Each individual is associated with one enterotype group	Each individual is associated with multiple co-abundance groups
Enterotypes are mostly defined by the most abundant genera. Normally *Bacteroides*, *Prevotella* and *Ruminococcus* (or another genera)	The definition of co-abundance groups is based on gradients of taxa and associations between taxa
The Enterotypes definition is rigorous	Stable associations defining co-abundance groups are yet to be finalized

Interest in these techniques has been huge with the formation of large multinational scientific consortiums such as the The Human Microbiome Project [10], MetaHIT (Metagenome of Human Intestinal Tract) [11] and the smaller ELDERMET project [12], as well as numerous labs around the world. These consortia have taken advantage of the new high-throughput technologies and have for the first time fully characterized the human microbiome in the gut and from other body sites.

The use of these techniques has illustrated the heterogeneity of the microbial populations in our gastrointestinal tracts with large inter-individual differences in the presence and absence of the bacterial species. Although some species are present in the majority of the population, these are in the minority in terms of the number of species that can be found in our gut. These rarer species are no less important for the well-being of the host. Species tend to co-occur and may be clustered into co-abundance groups (CAGs) [12] due to habit preference as defined by diet and cross-feeding events and the presence of bacteriocins, a type of bacterially produced antibacterial agent that is specific for a limited number or range of species. An alternative to the idea of CAGs are enterotypes. The idea of enterotypes predates that of CAGs and is different in a number of characteristics (table 1) [13]. Enterotype groups are distinct from one another and are often described as being similar to blood groups. Despite being controversial, the idea has become popular and has allowed researchers to categorize samples based on the dominant genera that represent microbial populations that have a substantial scope to modify the phenotype of the individual through production of metabolites and immunomodulatory effects.

Culture-Independent Microbiota of Older Persons

Microbial-Based Changes in the Elderly

Numerous studies from different geographical locations have attempted to characterize the microbiota of general healthy populations, and many have compared these with individuals carrying diseases, elderly, and even extreme elderly – individuals over 100 years of age. In 2001, Hopkins et al. [14] analyzed bacterial 16S rDNA sequences from children, adults, elderly, and *C. difficile*-infected geriatric patients from the UK. They revealed an overall decrease in bifidobacteria in elderly compared with adults, and a slight decrease in lactobacilli. There was no change in the *Bacteroides-Porphyromonas-Prevotella* group, contrary to their culture-based study 3 years later [7], discussed above. This group published again in 2004 [15], using real-time PCR on rDNA to compare bacteria from healthy elderly, hospitalized elderly, and elderly treated with antibiotics. The *Bacteroides-Prevotella* species were significantly less abundant in the hospitalized patients than in healthy elderly, whereas *Enterobacteriaceae*, *Clostridium butyricum* and *Enterococcus faecalis* were increased. Antibiotic-treated patients showed an increased abundance of *E. faecalis* compared with healthy elderly, but decreased abundances of *Bacteroides-Prevotella* group, *Desulfovibrio*, *Enterobacteriaceae*, *Faecalibacterium prausnitzii*, *C. butyricum* and *Ruminococcus albus*. *F. prausnitzii* has an anti-inflammatory affect, so reduced levels in antibiotic-treated patients may be associated with inflammaging.

Other studies have focused on other aspects of aging, such as frailty. A study on long-term care subjects in one elderly centre in The Netherlands [16] showed that an increase in frailty correlated with an increase in *Ruminococcus* and *Atopobium*, and a decrease in the *Bacteroides-Prevotella* group, the *Eubacterium rectale-Clostridium coccoides* cluster, *Lactobacillus* and *F. prausnitzii*. This frail microbial signature was similar to that found in the hospitalized subjects discussed by Bartosch et al. [15] and Claesson et al. [12].

Cultural Microbial Differences Observed in Different Age Groups

A European study of subjects from four different countries, France, Italy, Germany and Sweden, provides evidence of location-based differences [17]. No differences were observed between age groups from the French or Swedish cohorts. The *E. rectale-C. coccoides* group increased with age in the German population, but decreased with age in the Italian subjects, a decrease similar to the Dutch study by Bartosch et al. [15]. German adults had lower *Bacteroides-Prevotella* than adults from other countries, while Italian elderly had lower proportions than elderly from other countries. *F. prausnitzii* decreased with age in the Swedish and Italian subjects, but not the French or German subjects. The *Atopobium* cluster increased with age in German and Swedish populations, but not French and Italian. *Bifidobacterium* was lower in all elderly subjects than their corresponding adult cohorts; however, this was not significant. Italian subjects had significantly more bifidobacteria than other populations.

Claesson et al. [12] assessed the microbiota of a large cohort of Irish elderly, from four different residence locations – community, long-term care, rehabilitation and day-hospital. They reported an overall trend of increasing *Bacteroidetes* and decreasing *Firmicutes* from community to long-stay. Reduced abundances of *Coprococcus* and *Roseburia* were observed in long-stay subjects, while they had increases in *Parabacteroides*, *Eubacterium*, *Anaerotruncus*, *Lactonifactor* and *Coprobacillus*. This relatively large study was able to identify a number of microbial relationships between microbiota and frailty and other clinical factors while controlling for confounding factors such as diet, medication and even age.

There are few differences between young and healthy elderly subjects with the recurring associations with increased Proteobacteria and Bifidobactia. Biagi et al. [18] attempted to address this with centenarians. The centenarians tended to cluster separately from elderly and young, which did not differ. The mainly centenarian cluster had higher Proteobacteria and Bacilli, and lower *Clostridium* cluster XIVa, but no reduction in Bacteroidetes was observed.

These studies convey the large inter-individual differences in microbiota in a given population. They also provide us a view of some of the differences that can be seen between populations. These illuminate the issues with generalized views of microbiota, and remind us of some of the difficulties we will face when trying to increase longevity and quality of life. Somewhat personalized or community-based approaches may need to be considered in the future.

Diet as a Driver of Microbiota Variation in Older People

Factors Influencing the Gut Microbiota in Older Persons

There are a number of challenges facing the study of the role of diet as a modulator of gut microbiota variation in older persons including: (1) compositional inter-individual variability of the gut microbiota; (2) inter-individual variance in dietary intakes even among seemingly homogenous population groups, e.g. the elderly [12]; (3) variable effects of dietary intervention, dependent on the baseline microbiota; (4) the use of medical therapeutics, especially antibiotics [19]; (5) classification of an appropriate timeframe for dietary intervention and how to quantify its total microbiome effects.

The bacterial residents of the human gastrointestinal tract are capable of synthesizing a number of micronutrients including a number of B-group vitamins, vitamin K_2 isomers collectively known as menaquinones, and can aid calcium absorption, and regulate appetite and insulin release (fig. 1). In fact, long-term dietary choices can greatly influence the gut microbiota population, while nutritional status has also been shown to impact the composition of the gut microbiota and its functionality [20]. Dietary choice and variety are also influenced by disease which can affect the composition, diversity and metabolic capacity of the gut microbiota and have important implications for therapies aimed at modulating the large intestinal microbiota. The ELDERMET study illustrated the negative clinical implications of reduced dietary and microbiota diversity [12]. Targeted dietary interventions using prebiotics, probiotics

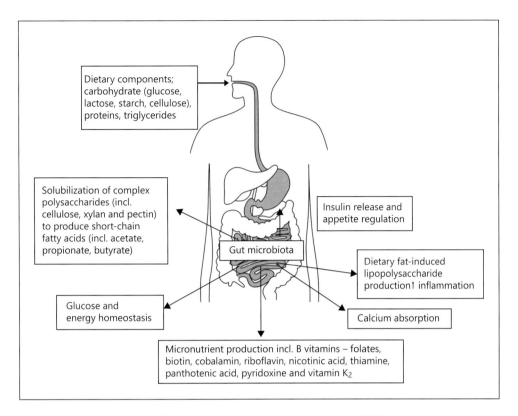

Fig. 1. Interaction between diet, gut microbiota and nutritional status [37].

or combinations (i.e. synbiotics) may counteract negative age-related changes, and address the detrimental effects of long-term broad-spectrum antibiotic therapy on the gut. Limited feeding trials show promising results with these supplements.

The Impact of Diet on Microbiota Variation

Diet can have a marked impact on the gut environment, including transit time and pH. Changing the intakes of the three main macronutrients (carbohydrates, protein and fats) can significantly affect the composition of the microbiota. In a study of the effects of long-term diet on the microbiota composition, Wu et al. [21] found that high dietary intakes of animal products were associated with *Bacteroides*, while a greater intake of plant material promoted the *Prevotella* genus. Similarly, the ELDERMET study found an abundance of bacterial species from the *Prevotella* genus among healthy elderly individuals [12]. A dietary intervention trial investigating the effect of different non-digestible carbohydrate (resistant starch and non-starch polysaccharide) on overweight individuals, revealed the adaptive nature of the microbiota [22]. Plant polysaccharides, excluding starches, are key structural and biological components of the cell membrane and are collectively known as non-starch polysaccharides. These structural components cannot be hydrolyzed by the endogenous enzymes of humans. Instead, a complex mutual dependence has developed between the mammalian host and symbiotic gut microorganisms that possess the ability to access this additional energy source which can contribute up to 10% of the body's daily energy requirements. In addition, extensive cross-feeding occurs in the colon between the primary degraders of complex substrates and other bacterial species that depend on their fermentation and partial degradation products from dietary substrates.

The importance of diet in modifying the gut microbiota in elderly populations is becoming increasingly apparent. A review by Woodmansey et al. [23] suggests a reduced and enhanced complement of carbohydrate and protein metabolizing bacteria in old age, respectively. Interestingly, it is thought increased proteolytic activity in the gut, caused by increased *Bacteroides* and *Clostridium* species is responsible for putrefaction and production of harmful ammonia, amines, phenols and indoles, while high fat feeding has been associated with increased systemic endotoxin production and low-grade inflammation in animal models [24]. High-fat, high-protein, low-complex carbohydrate diets promote the production of colonic residues that promote microbial production of potentially carcinogenic byproducts [25]. It is thought that the microbiota mediates the effect of diet on colon cancer risk by regulating the generation of butyrate, folate, and biotin, molecules known to play a key role in the regulation of epithelial proliferation. However, unlike high protein and fat intakes, carbohydrate increases gut microbial diversity [21], which has been associated with several health benefits in a number of population groups, including older populations [12]. Long-term high dietary fibre consumption has been associated with increased bacterial genes for cellulose and xylan hydrolysis and increased production of short-chain fatty acids [26], an energy source for epithelial cells which have been shown to have anti-inflammatory properties and may be particularly beneficial to aging population groups.

Dietary Interventions

The majority of dietary intervention data has focused on the bifidogenic role of prebiotics in human subjects, which are defined as non-digestible carbohydrates, assumed to infer benefits for the host health by stimulation of a protective intestinal microbiota. The majority of research in this area has focused on the modulatory effects of single carbohydrate fractions, predominantly long- and short-chain carbohydrates which escape digestion in the upper gastrointestinal tract. These include fructo-oligosaccharides (FOS), galacto-oligosaccharides (GOS), resistant starches and prebiotics such as inulin and oligofructose. They are then fermented in the colon and selectively stimulate the growth of bifidobacteria, which are known to exist at mark-

Table 2. Prebiotic intervention studies in elderly populations

Study design	Subjects	Age, years	Intervention	Intervention period	Dose, g/day	Result	Ref.
Randomized, double-blind, parallel study	10 (inulin) 15 (lactose)	68–89	inulin or lactose	19 days	20 (day 1–8) 40 (day 12–19)	↑ Bifidobacteria ↓ Enterobacteria ↓ Enterococci	[34]
Single-treatment study	37	77–91	FOS	3 weeks	8	↑ Bifidobacteria	[35]
Single-treatment study	12	67–71	scFOS	4 weeks	8	↑ Bifidobacteria	[36]
Randomized, double-blind, placebo-controlled, crossover study	44	64–79	B-GOS, placebo	10 weeks	5.5	↑ Bifidobacteria ↑ *Lactobacillus-Enterococcus* spp. ↑ *C. coccoides-E. rectale* group ↓ *C. histolyticum* group ↓ *E. coli* ↓ *Desulfovibrio* spp.	[28]
Randomized, double-blind, placebo-controlled, crossover study	37	50–81	GOS, placebo	3 weeks	4	↑ Bifidobacteria ↑ butyrate	[27]

scFOS = Short-chain fructo-oligosaccharides; B-GOS = prebiotic GOS mixture.

edly reduced levels in old age (table 2). Studies have shown that prebiotics affect the immune system as a direct or indirect result of changes in the gut microbiota profile or in the fermentation potential. In addition, GOS and FOS have been attributed to functional claims related to the bioavailability of minerals, lipid metabolism and regulation of bowel habits.

Prebiotic Intervention Studies in Elderly Populations

A limited number of intervention trials have been conducted to determine the microbiota-modulatory effects of certain prebiotic ingredients. Walton et al. [27] conducted a study with thirty-nine, 50- to 81-year-olds supplemented with 8 g GOS or placebo for 6 weeks and found significant increases in bifidobacteria and butyrate. GOS-induced bifidogenesis was also reported by Vulevic et al. [28], who focused on the immunomodulatory effects of a mixed GOS preparation over a 10-week period, resulting in increased phagocytosis and an anti-inflammatory cytokine profile. However, the inter-individual variation in gut microbiota suggests the response of microbial communities to prebiotic supplementation/dietary modulation may also vary. In fact, certain study volunteers have been reported as 'non-responders', indicating the influence of the initial composition of the gut microbiota.

Human studies have also investigated the effects of arabinoxylan-oligosaccharide supplementation, which have been reported to impact the protein/carbohydrate fermentation balance in the large intestine and thus affect the generation of potentially toxic metabolites in the colon. Supplementation trials have been reported to increase total bacterial populations including *Bifidobacterium*, faecal butyrate concentrations [27] and fermentation activity. Further investigation into the positive modulatory effects of other promising functional ingredients on microbiota composition in elderly cohorts is warranted.

Prospectus, Knowledge Gaps and Required Studies

Diet-Microbiota Interactions

It is clear from the preceding text that diet is a major driver of microbiota variation in the elderly (and other age groups). The large differences in the trending microbiota composition changes in European subjects in different countries measured by the CROWNALIFE study [17] are almost certainly due to factors including diet. As well as differences in the methods employed in culture-dependent studies, the difficulty in identifying unifying trends in microbiota composition change upon aging, as reviewed by Woodmansey [23], is also probably due in large part to diet influences. It is therefore necessary to conduct large longitudinal microbiota measurements in well-phenotyped individuals who consume a carefully measured diet, as well as to preform controlled dietary interventions in subjects whose microbiome has been analyzed at baseline. The NuAge project, funded by the European Commission (www.nu-age.eu), will go some way to tackling the latter aspiration because one of the objectives is to conduct a dietary intervention in 1,250 subjects across 5 geographically spread-out European cities, whereby the participants will be switched to a Mediterranean diet. Extensive physical and clinical measurements will then seek to correlate any changes recorded with alterations in the microbiome, inflammasome and peripheral blood lymphocyte epigenome in the subjects.

Diet-Microbiota-Medication Interactions

Even in healthier subjects, consumption of medication, often multiple types, is common in the elderly. However, there have been no dedicated studies of how the gut microbiota affects or is affected by multiple medications. At least 30 drugs are already known to be modified by the microbiota [29], and there is good evidence that the activity of the microbiota can affect the activity or bioavailability of drugs [30]. Coupled with dietary effects, this could present a complicated environmental modifier with particular relevance in the elderly. To unravel the interactions, strongly powered studies will be required.

Ecosystem Management, Microbiota Restoration and Replacement

The term 'dysbiosis' is commonly used to convey an altered microbiota, but the difficulty of defining a core microbiota makes it very challenging to describe when dysbiosis has occurred. Despite this, what has certainly become clear from recent studies is that a low-diversity microbiota is a common feature of subjects undergoing obesity, inflammatory bowel disease [31] and accelerated aging-related health loss [12]. As noted above, our studies of increased frailty and health loss in long-term residential care subjects were characterized by a lower diversity microbiota and lower gene counts in shot-gun metagenome data [12]. While dietary adjustment would be the simplest way to restore microbiota diversity, it may prove practically difficult to implement because of physiological or financial barriers. Furthermore, in older subjects, there is a risk that the missing elements of the microbiota cannot be restored by diet alone. Faecal microbiota transplants have proved effective for treating *C. difficile* in humans [32] and for restoring glucose sensitivity in patients with metabolic syndrome [33]. Transplantation of the entire microbiota between humans poses residual safety concerns, but development of artificial consortia based on defined cultured microbes offers the prospect of a clean reproducible alternative. In the case of older subjects, one can imagine there being individuals who have lost the entire community of microorganisms capable of hydrolyzing or metabolizing certain dietary ingredients, such as resistant starch or other complex polysaccharide. So-called 'bacteriotherapy' may also be the only way of restoring desired microbiota elements to these individuals.

References

1 de Vos WM, de Vos EAJ: Role of the intestinal microbiome in health and disease: from correlation to causation. Nutr Rev 2012;70:S45–S56.

2 O'Toole PW, Claesson MJ: Gut microbiota: changes throughout the lifespan from infancy to elderly. Int Dairy J 2010;20:281–291.

3 Clemente JC, et al: The impact of the gut microbiota on human health: an integrative view. Cell 2012;148:1258–1270.

4 Tiihonen K, Ouwehand AC, Rautonen N: Human intestinal microbiota and healthy ageing. Ageing Res Rev 2010;9:107–116.

5 Benno Y, et al: Comparison of fecal microflora of elderly persons in rural and urban areas of Japan. Appl Environ Microbiol 1989;55:1100–1105.

6 Hopkins MJ, Macfarlane GT: Changes in predominant bacterial populations in human faeces with age and with *Clostridium difficile* infection. J Med Microbiol 2002;51:448–454.

7 Woodmansey EJ, et al: Comparison of compositions and metabolic activities of fecal microbiotas in young adults and in antibiotic-treated and non-antibiotic-treated elderly subjects. Appl Environ Microbiol 2004;70:6113–6122.

8 Rajilić-Stojanović M, et al: Development and application of the human intestinal tract chip, a phylogenetic microarray: analysis of universally conserved phylotypes in the abundant microbiota of young and elderly adults. Environ Microbiol 2009;11:1736–1751.

9 Margulies M, et al: Genome sequencing in microfabricated high-density picolitre reactors. Nature 2005;437:376–380.

10 Human Microbiome Jumpstart Reference Strains Consortium, Nelson KE, Weinstock GM, Highlander SK, et al: A catalog of reference genomes from the human microbiome. Science 2010;328:994–999.

11 MetaHIT Consortium: MetaHIT draft bacterial genomes at the Sanger Institute, 2010. http://www.sanger.ac.uk/resources/downloads/bacteria/metahit/.

12 Claesson MJ, et al: Gut microbiota composition correlates with diet and health in the elderly. Nature 2012;488:178–184.

13 Arumugam M, et al: Enterotypes of the human gut microbiome. Nature 2011;473:174–180.

14 Hopkins MJ, Sharp R, Macfarlane GT: Age and disease related changes in intestinal bacterial populations assessed by cell culture, 16S rRNA abundance, and community cellular fatty acid profiles. Gut 2001; 48:198–205.

15 Bartosch S, et al: Characterization of bacterial communities in feces from healthy elderly volunteers and hospitalized elderly patients by using real-time PCR and effects of antibiotic treatment on the fecal microbiota. Appl Environ Microbiol 2004;70:3575–3581.

16 van Tongeren SP, et al: Fecal microbiota composition and frailty. Appl Environ Microbiol 2005;71: 6438–6442.

17 Mueller S, et al: Differences in fecal microbiota in different European study populations in relation to age, gender, and country: a cross-sectional study. Appl Environ Microbiol 2006;72:1027–1033.

18 Biagi E, et al: Through ageing, and beyond: gut microbiota and inflammatory status in seniors and centenarians. PLoS One 2010;5:e10667.

19 O'Sullivan Ó, et al: Alterations in intestinal microbiota of elderly Irish subjects post-antibiotic therapy. J Antimicrob Chemother 2013;68:214–221.

20 Flint HJ: The impact of nutrition on the human microbiome. Nutr Rev 2012;70:S10–S13.

21 Wu GD, et al: Linking long-term dietary patterns with gut microbial enterotypes. Science 2011;334: 105–108.

22 Walker AW, et al: Dominant and diet-responsive groups of bacteria within the human colonic microbiota. ISME J 2011;5:220–230.

23 Woodmansey EJ: Intestinal bacteria and ageing. J Appl Microbiol 2007;102:1178–1186.

24 Cani PD, et al: Involvement of gut microbiota in the development of low-grade inflammation and type 2 diabetes associated with obesity. Gut Microbes 2012; 3:279–288.

25 Russell WR, et al: High-protein, reduced-carbohydrate weight-loss diets promote metabolite profiles likely to be detrimental to colonic health. Am J Clin Nutr 2011;93:1062–1072.

26 De Filippo C, et al: Impact of diet in shaping gut microbiota revealed by a comparative study in children from Europe and rural Africa. Proc Natl Acad Sci 2010;107:14691–14696.

27 Walton GE, et al: A randomised crossover study investigating the effects of galacto-oligosaccharides on the faecal microbiota in men and women over 50 years of age. Br J Nutr 2012;107:1466–1475.

28 Vulevic J, et al: Modulation of the fecal microflora profile and immune function by a novel trans-galactooligosaccharide mixture (B-GOS) in healthy elderly volunteers. Am J Clin Nutr 2008;88:1438–1446.

29 Sousa T, et al: The gastrointestinal microbiota as a site for the biotransformation of drugs. Int J Pharm 2008;363:1–25.

30 Wilson ID, Nicholson JK: The role of gut microbiota in drug response. Curr Pharm Design 2009;15:1519–1523.

31 Manichanh C, et al: The gut microbiota in IBD. Nat Rev Gastroenterol Hepatol 2012;9:599–608.

32 van Nood E, et al: Duodenal infusion of donor feces for recurrent Clostridium difficile. N Engl J Med 2013;368:407–415.

33 Vrieze A, et al: Transfer of intestinal microbiota from lean donors increases insulin sensitivity in individuals with metabolic syndrome. Gastroenterology 2012;143:913–916.e7.

34 Kleessen B, et al: Effects of inulin and lactose on fecal microflora, microbial activity, and bowel habit in elderly constipated persons. Am J Clin Nutr 1997;65: 1397–1402.

35 Guigoz Y, Lauque S, Vellas BJ: Identifying the elderly at risk for malnutrition: The Mini Nutritional Assessment. Clin Geriatr Med 2002;18:737–757.

36 Bouhnik Y, et al: Four-week short chain fructo-oligosaccharides ingestion leads to increasing fecal bifidobacteria and cholesterol excretion in healthy elderly volunteers. Nutr J 2007;6:42.

37 O'Connor EM: The role of gut microbiota in nutritional status. Curr Opin Clin Nutr Metab Care 2013; 16:509–516.

Paul W. O'Toole
School of Microbiology UCC
Room 447 Food Science Building
University College Cork
Western Road
Cork (Ireland)
E-Mail pwotoole@ucc.ie

Yashin AI, Jazwinski SM (eds): Aging and Health – A Systems Biology Perspective.
Interdiscipl Top Gerontol. Basel, Karger, 2015, vol 40, pp 155–176 (DOI: 10.1159/000364981)

Systems Biology Approaches in Aging Research

Anuradha Chauhan[b] · Ulf W. Liebal[b, f] · Julio Vera[a] · Simone Baltrusch[c] ·
Christian Junghanß[d] · Markus Tiedge[c] · Georg Fuellen[e] ·
Olaf Wolkenhauer[b] · Rüdiger Köhling[f]

[a]Laboratory of Systems Tumor Immunology, Department of Dermatology, University Hospital Erlangen and
Friedrich-Alexander Universität Erlangen, Erlangen, [b]Department of Systems Biology and Bioinformatics,
Institute of Computer Science, University of Rostock, [c]Institute of Medical Biochemistry, [d]Department of
Hematology, [e]Institute for Biostatistics and Informatics in Medicine and Ageing Research of Physiology,
[f]Institute of Physiology, Rostock University Medicine, Rostock, Germany

Abstract

Aging is a systemic process which progressively manifests itself at multiple levels of structural and functional organization from molecular reactions and cell-cell interactions in tissues to the physiology of an entire organ. There is ever increasing data on biomedical relevant network interactions for the aging process at different scales of time and space. To connect the aging process at different structural, temporal and spatial scales, extensive systems biological approaches need to be deployed. Systems biological approaches can not only systematically handle the large-scale datasets (like high-throughput data) and the complexity of interactions (feedback loops, cross talk), but also can delve into nonlinear behaviors exhibited by several biological processes which are beyond intuitive reasoning. Several public-funded agencies have identified the synergistic role of systems biology in aging research. Using one of the notable public-funded programs (GERONTOSYS), we discuss how systems biological approaches are helping the scientists to find new frontiers in aging research. We elaborate on some systems biological approaches deployed in one of the projects of the consortium (ROSage). The systems biology field in aging research is at its infancy. It is open to adapt existing systems biological methodologies from other research fields and devise new aging-specific systems biological methodologies.

© 2015 S. Karger AG, Basel

The Emergence of Systems Biology 'Flavors'

The Conceptual Workflows of Systems Biology

Systems biology is a multidisciplinary approach that has been developed over the last 15 years, and which attempts to elucidate the structural and functional organization of

---

A.C., U.W.L. and J.V. contributed equally to this work.

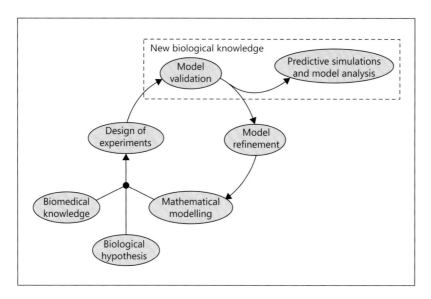

Fig. 1. General structure of the canonical systems biology method. The procedure includes: construction of the regulatory map for the system under investigation using existing biomedical information; set up of mathematical model representing interesting part of the map; characterization of the model equations calibration using quantitative experimental data; model assessment and refinement; model validation through additional experiments, and use of the model to make predictions on the basic features of the system, which are tested with additional experiments.

biological systems at the intracellular level as well as the interaction between cells and their microenvironment [1]. What makes the methodology special is that questions concerning the structure and function of the biological systems are investigated using quantitative experimental data, which are analyzed through the use of sophisticated mathematical and computational tools, such as advanced statistics, data mining and mathematical modelling. The rationale for the methodology is to apprehend the complexity of biological systems, to derive nonintuitive hypotheses about their functioning, to help designing and formulating new experiments able to prove these hypotheses, and, ultimately, to develop computational tools with predictive ability in a biomedical focus.

In its original conception, the method implied iterative cycles of hypothesis formulation, design of experiments, and integration of the obtained data using advanced mathematical tools like modelling and simulation [1, 2]. The method included the following steps (fig. 1):

(a) Relevant biomedical knowledge is retrieved from publications, databases and public repositories of biological data using computational tools, text mining and manual curation. The information is used to construct a graphical depiction of the biochemical system (i.e. the network), its compounds (mRNAs, microRNAs, proteins, small molecules, protein complexes…), as well as relevant interactions. This graphical depiction is often called 'regulatory map' of the system and contains the up-to-date biological, biomedical, and sometimes clinical, knowledge on the investigated system.

(b) The regulatory map itself is a resource that can be analyzed and mined using computational tools from network biology. This helps organizing the network in modules relevant to the investigated problem, or finding unexpected connections between pathways thought to be independent (i.e. cross talk).

(c) Critical parts of the regulatory network are translated into a mathematical model, consisting of nothing but a set of mathematical equations that encode the knowledge and hypothesis about the systems investigated. Under some circumstances, the analysis of this preliminary model can yield already useful information. But in most of the cases, the model has to be better characterized using quantitative data generated in customized experiments. Usually, one proceeds in iterative cycles of mathematical model derivation, integration of the experimental data in the model and reformulation of model equations in case the agreement between the experimental data and the model simulations is not satisfactory.

(d) The output of this process is a computational tool, the *calibrated* mathematical model, which has predictive abilities. This means that model simulations can be used to predict the behavior of the network under experimental scenarios not yet tested. In fact, many examples of validated mathematical model are available in the literature, for example to predict drug targets [3] or chemoresistance [4], and to identify diagnosis biomarkers [5] for cancer and other human diseases.

Systems Biology Comes in Several 'Flavors'

While in the last years several new communities of researchers (with very different scientific background) have joined the biomedical sciences in an attempt to break down prevalent diseases like cancer and diabetes, the philosophy behind the systems biology approach has been expanded, and other workflows can be found in the recent literature under this epigraph. The common point among all of them is the intensive use of advanced mathematical and computational tools to analyze, dissect, mine and make sense out of large sets of quantitative experimental data. We can distinguish between the omics approach, reconstruction and simulation of large biological networks, elucidation of the complexity emerging from nonlinear network motifs, and multi-scale data integration and modelling.

The Omics Approach
In an increasing amount of subfields, researchers face the problem of analyzing and integrating massive amounts of high-throughput biological or biomedical data. In this case, it has become very popular to make use of advanced statistical techniques in an attempt to identify global expression patterns. In cancer and other multifactorial diseases, large amounts of high-throughput and clinical data obtained from large cohorts of patients are analyzed using this strategy, aiming at 'fishing' unknown and unexpected molecular mechanisms triggering the disease. Or even more important,

obtaining reliable sets of biomarkers (the so-called gene signatures), able to accurately stratify populations of patients in a predictive fashion, making possible to make more precise the diagnosis and foresee the patient response to therapies [6].

Reconstruction and Simulation of Large Biological Networks
In the last years, the certainty emerged that intracellular biochemical processes can rarely be compartmentalized into small pathways, especially when considering biomedical questions. The new pictures of many diseases show massive and complex biochemical networks, composed of coding genes, small and long noncoding RNAs, metabolites, small molecules and interacting proteins, all of them interconnected by a myriad of different kinds of interactions and molecular modifications [7]. The natural way to characterize these systems is the use of multi-level sets of in vitro or in vivo high-throughput data. Under these conditions, human intuition or basic data analysis techniques are clearly inadequate and limited tools, and hence advanced methods for data analysis, network reconstruction and simulation are required, which have been used with remarkable success in the last years [4, 8].

Elucidation of the Complexity Emerging from Nonlinear Network Motifs
If the size of biomedical relevant networks and the multiplicity of high-throughput data sources necessary to characterize them was already a challenge big enough, we have found that these networks are extensively enriched in nonlinear network motifs. We mean systems of interactions between network compounds displaying the structure, and therefore the behavior, of structures like positive and negative feedback loops, but also coherent and incoherent feed-forward loops, or network hubs (see fig. 2) [7, 9, 10]. Even more challenging, rather than isolated, these motifs very often overlap and cross talk, resulting in a structural and functional complexity that is far beyond the understanding of our mind [11]. However, feedback loops have been extensively investigated for decades by engineers and physicists, who have developed a vast framework of analysis based on the use of mathematical modelling. This framework has been largely exploited in systems biology [12, 13].

Multi-Scale Data Integration and Modelling
We know now that in order to apprehend the complexity of many biological phenomena, we have to go beyond the scale of the intracellular biochemical processes, and consider the communication between cells and their microenvironment, the dynamics of tissue organization and replenishment, and even the influence of the whole-body processes. When integrating quantitative biological data with different structural, spatial and temporal scales of these biological phenomena (e.g. intracellular-omics data, live microscopy and systemic blood markers), the so-called multi-scale mathematical modelling has emerged as a tool to make sense out of this complexity.

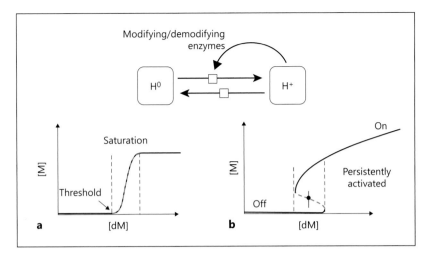

Fig. 2. Biochemical networks are enriched in nonlinear motifs, like positive feedback loops (left), a regulatory structure in which the activation of a signaling event positively regulates a signaling process upstream of the pathway. Positive feedback loops can induce signal amplification, ultrasensitivity to input signal, but also in some special cases bistability (right). **a** Ultrasensitivity enables a cellular system to transform graded input signals at a certain threshold into discrete all-or-none output. **b** In systems showing bistability, for some given experimental regime, small perturbations in the setup may induce totally different fates for the system, inducing for example quick and irreversible signal termination or persistent activation.

Why Systems Biology in Aging?

If we integrate the definition of systems biology and the different subfields discussed in previous sections with aging research, then we can conclude that aging research appears to be an ideal research field to deploy extensive systems biology efforts.

Aging Phenotypes Are Associated with Large and Complex Networks

Contrary to most of the other fields in biomedicine, the notion of biochemical networks and their dysfunction as drivers for the emergence of aging has been firmly established for decades. Back in 1994, Kowald and Kirkwood hypothesized that multiple biological processes, some of them involving the progressive dysfunction of critical metabolic and repair pathways, work in parallel and probably cross talk during the emergence of aging [16, 17]. To test the consequences of their hypothesis, they derived, simulated and analyzed a mathematical model, which integrates several existing theories about the triggers of aging at the cellular level, including the so-called free radical theory and the protein error theory. The model results were in agreement with the caloric restriction hypothesis and its role in reducing free radical production, protein damage and hence extending life span.

Franceschi and coworkers further extended the network theory of aging by including the role of the immune system and the inflammatory signals in it [18]. In their opinion, an age-related decline in the ability of the tissues to mitigate the effects of environmental and internal stress, probably connected to the progressive dysfunction of the damage repair systems hypothesized by Kirkwood and Kowald, triggers a progressive systemic proinflammatory phenotype. This proinflammatory phenotype negatively influences body performance and contributes to the emergence of many aging-associated diseases like Alzheimer's disease, osteoporosis and diabetes.

To a certain extent, Franceschi's hypothesis of the so-called 'inflamm-aging' supports the notion that aging is the consequence of the progressive deregulation of a 'network of networks'. In line with this, De Magalhaes and Toussaint developed the first curated database of genes related to human aging [19]. The information contained in the database was further used to construct the first large proteomic network map of human aging. Interestingly, when they analyzed the obtained network, they found that many of the genes in their aging network are commonly associated with the genetics of development. This work has been continued by other groups who have worked in establishing the topology of this aging 'network of networks'. To mention an additional example, Xue and collaborators combined information on protein-protein interactions and gene expression data during fruit fly life span to construct and modularize a network of interacting proteins playing a role in aging [20]. Interestingly, their results pinpoint to a reduced number of network modules as critical mediators of aging. But they also suggest that the structure of the obtained network was such that dysfunction in few critical regulatory nodes, which connect some of these aging modules, may be behind what they call 'the molecular basis for the stochastic nature of aging'.

Motifs of Nonlinear Dynamics in Aging Networks

The aging network is enriched in motifs that govern the emergence of aging in a timely and spatially nonlinear fashion. Further investigation into the network and the pathways guilty of mediating aging emergence found that, to the surprise of many researchers, they are enriched in motifs like feedback and -forward loops [21, 22]. Thus, their likely nonlinear dynamics are in clear contradiction to the vision of aging as a progressive process. Furthermore, some results support the idea that at least some of the critical events triggering aging emergence may not behave linearly [18]. As Kitano, Kirkwood and others mention, the motivation behind the existence of large biochemical networks enriched in nonlinear motifs may be to provide robustness against

Chauhan · Liebal · Vera · Baltrusch · Junghanß · Tiedge · Fuellen · Wolkenhauer · Köhling

perturbation, and the loss of this robust performance due to pathway dysfunction may be the basis of several mechanisms involved in aging [23–25].

Further examples illustrate the effect of nonlinear regulatory circuits involved in aging-related phenotypes. Passos and collaborators found a delayed feedback loop system involving a long signaling circuit, by which long-term activation of p21 and reactive oxygen species (ROS) production and subsequent cell senescence can be activated upon a sufficiently long-lasting DNA damage response [26]. In line with this, Lai and collaborators show that p21 can be regulated in a cooperative manner by multiple miRNAs, which allows for a context-dependent regulation of p21 signaling in several aging-associated contexts. Kriete and collaborators constructed a regulatory network, which included most of the central cellular mechanisms involved in aging, and showed that most of them were connected and cross-talked via a number of overlapping positive and negative feedback loops [22]. When translated into a mathematical model, their model simulations showed that dysfunction of key, positive-feedback loop-regulated processes in the cellular energy metabolism may lead to enhanced and accelerated cellular damage. Furthermore, they showed that a number of negative-feedback loop systems involved in the NFκB and mTOR signaling have the ability to alleviate cellular damage by regulating processes that are essential for cell survival, such as mitochondrial respiration or biosynthesis, and their deregulation may enhance aging phenotypes.

Aging as a Systemic Phenomenon

Aging is not a cellular phenotype, but it manifests at multiple levels of organization in the human body. It is even intuitive to say that aging as a phenomenon evades the frontiers of the intracellular machinery, and involves complex interactions between the cell and its microenvironment, the role of cell dynamics in tissue organization and replenishment, but also higher-scale processes connecting the functioning of organs with systemic, whole-body like processes. The emerging multi-scale system (cell damage → tissue/organ dysfunction → whole body deregulation) is enriched in equally multi-scale feedback loops, which connect the progressive cellular dysfunction with systemic stress signals, and the way back (systemic stress signals → tissue/organ dysfunction → cell damage [18, 24]. In line with this perception, the ultimate aim of aging research is to create a framework to connect experimental evidence of aging progression emerging at different structural, spatial and temporal scales, similar to what has been done in cancer [14].

The use of this approach in aging is still in its infancy, but we still can find some notable examples. Hoehme and collaborators derived and characterized using multiple sources of data a multi-scale mathematical model accounting for the liver regeneration after damage in mice [27]. Remarkably, the model was able to predict novel essential mechanisms for liver regeneration, a process that must be critical to impede aging-associated deterioration of the liver. Van Leeuwen and coworkers made use of mathematical modelling to interconnect biological processes happening at different organizational scales within the body, which connect caloric restriction and the dynamics of oxidative cellular damage, metabolism, body weight change and ultimately life span extension [28, 29]. The obtained mathematical model was able to reproduce growth and survival data on mice exposed to different food levels and supported the caloric restriction hypothesis. McAuley and collaborators developed a whole-body mathematical model of age-related deregulation of the cholesterol metabolism [30]; for its characterization, they used multifactorial data, accounting for the dynamics of the system from both the molecular and the physiological perspectives. Model simulations and analysis were used to investigate the actual relevance of known molecular mechanisms in the age-associated deregulation of cholesterol.

From Understanding to Manipulation

When thinking about fighting aging, an engineering approach can be used. This idea is behind one of the most controversial theories in aging research, the so-called SENS hypothesis postulated by Aubrey de Grey [31, 32]. This hypothesis assumes that aging emerges due to the progressive accumulation of different kinds of damage during life span, which act in a potentially synergistic manner. These sources of systemic damage include mutations and epigenetic modifications in critical genes, stress-induced mutations in the mitochondrial genome, as well as intracellular and extracellular accumulation of different sorts of cellular debris, which are the basis of many aging-linked diseases. The concept is quite in agreement with Kowald and Kirkwood's Network Theory of Aging. However, the disputed point of de Grey's theory is that it proposes an engineering-like approach to fight the emergence of aging, which neglects the deep understanding of aging's origin and focuses on the design of biomedical therapies aimed at mitigating, preventing or repairing the sources of these cumulative organic damage that defines aging.

If at some point in the near future, the idea behind the SENS hypothesis becomes prevalent, then systems biology-like approaches would become an even more integrative part of aging research: system biology is in fact inspired by the methodology used by physicists and engineers to design and optimize complex technological devices, whose behavior is governed by nested regulatory circuits, especially feedback loops, as it is in the case of the network of networks whose dysfunction causes the emergence of aging.

Aging Research Projects Using the Systems Biology Approach: The GerontoSys Initiative

Having in mind the motivations discussed in the previous section, it is not a surprise that worldwide several public funding agencies have identified the synergistic role of systems biology in aging research and promoted in the very last years a number of

Chauhan · Liebal · Vera · Baltrusch · Junghanß · Tiedge · Fuellen · Wolkenhauer · Köhling

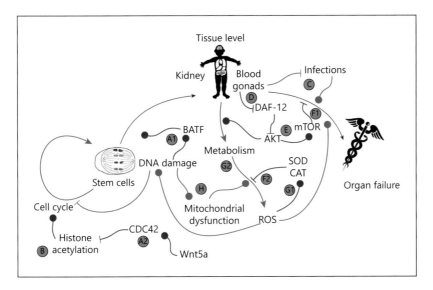

Fig. 3. Intracellular biochemical systems, tissues and phenotypes related to aging emergence and some systems biology projects addressing them. Legend of projects: A = SyStaR, B = MAGE, C = Gerontoshield and Primage, D = SyBACol, E = NephAge, F = GerontoMitosys, G = JenAge, H = ROSage.

projects making use of systems biology in aging research. A remarkable one is the Gerontosys initiative funded by the German Federal Ministry of Education and Research, which is being funded for the period 2009–2016, with a global budget of about EUR 32 million. Figure 3 summarizes the majority of projects funded in the second call. In the following, we examine some of these projects to illustrate the overall structure of an aging systems biology project and how most of them were conceived as rather large interdisciplinary research networks. Finally, we will focus on our own project, the ROSAge project, to provide an example for the initial conception, the work plan for the integration of different disciplines, and its development, but also the particular difficulties that one faces when approaching aging from a systemic view.

The Hematopoietic System Is an Aging Research Hub

Several projects in the Gerontosys initiative focus on aging in the hematopoietic system. Some of them examine the aging effects in hematopoietic stem cells (HSC), whereas others focus on aging-associated defects of the immune system. SyStaR (Molecular Systems Biology of Impaired Stem Cell Function in Regeneration during Aging; A in fig. 3) and MAGE (Model of the Aging Epigenome; B in fig. 3) are the projects with stem cell focus. Aging of the immune system is covered by Primage (Protective Immunity in Aging) and GerontoShield (C in fig. 3).

To mention some of the insights emerging from the systems biology approach used in these projects, the researchers involved in the SyStaR project could prove that HSC differentiation in response to DNA damage is a protection strategy which operated through the BATF (basic leucine zipper transcription factor-ATF like) pathway, and can reduce the burden of mutations in the stem cell niche (A1 in fig. 3). Interestingly, Wang et al. (2012) [33] corroborated the experimental findings theoretically by encoding the cellular processes of proliferation, differentiation and apoptosis into a mathematical model. The model simulations ruled out the role of apoptosis and strengthened the conclusion that differentiation is the driving force behind the reduction of the HSC pool in their niche, and not apoptosis. Moreover, the modelling and experimental results can explain the uneven differentiation levels between lymphoid and myeloid competent HSCs. The mechanistic interpretation emerging out of the systems biology analysis performed is that the increase in the BATF level stimulates predominantly the differentiation of lymphoid-competent HSCs, thereby draining their stem cell pool [33].

Histone Modifications and Stem Cell Fate

It is known that epigenetic histone modifications have a big impact on stem-cell-based organ regeneration and hence on the emergence of aging-associated organ deterioration. The team of the SyStar project could experimentally show epigenetic histone deacetylation via activation of Wnt5a and CDC42 signaling, which ultimately contributed to the cellular stem cell aging (A2 in fig. 3). In line with this, the effects of similar epigenetic changes on aging were investigated via mathematical modelling and simulation in the MAGE project [34, 35] (B in fig. 3). Interestingly, epigenetic modifications are processes inviting systems biological interpretations as they comprise several nonlinear network motifs like feedback loops [36], which are an integral part of engineering approaches, as discussed in the section Systems Biology Comes in Several 'Flavors'. Precisely, a positive feedback loop is generated when modified histones stimulate recruitment of modification enzymes and can lead either to a rapid increase of histone modifications (H^+), or to a second steady state with low modifications (H^0; see fig. 2) [36]. Modelling results indicate that this switch from high to low modifications can be triggered by the dilution of histone modifications during DNA replication [34]. Ultimately, this process leads to impaired stem-cell function, loss of stem cell pools and hence the emergence of aging-associated phenotypes.

Given the general importance of histone modifications in the aging process of HSCs, Przybilla and collaborators further performed an in silico simulation of the effect of histone and DNA modification rates on transcriptional activities of quiescent cells in the stem cell niches. They found that fluctuations in this DNA modification process could rejuvenate aged HSC and further activate their proliferation and differentiation capacity [35]. Genes of HSC become silenced during proliferation because of the dilution of histone methylation, and this silencing induces age-related phenotypes and reduces the overall the capacity of proliferation and differentiation. Because of the reduced proliferation, histone modification can reactivate silenced genes leading to stem cell rejuvenation. In line with this, Przybilla and coworkers' simulations indicate that an efficient DNA demethylation activity is necessary to enable sufficient plasticity of histone modifications [35]. Overall, the results emerging from the use of a systems biology workflow in the MAGE project suggest that aged cells with low proliferation rate accumulate in the HSC niche, and young active cells are pushed to differentiate, thus keeping the age of differentiated cells young.

Understanding mTOR Activation through Computational Modelling

mTOR is a central signal transducer of the nutritional cell status and growth signals [37]. In particular, organs with high metabolic activity, like liver, muscles and adipose tissue, and kidney are sensitive to mTOR signaling [37, 38]. Kidney diseases are among the leading ailments in the elderly, and the NephAge project was conceived to investigate the involvement of aging in them via a systemic approach (E in fig. 3). One of the starting points is that dynamics of the activation of the protein complex mTORC2 is insufficiently understood. Thus, the NephAge consortium ran an iterative cycle of hypothesis formulation, mathematical modelling analysis and experimental validation in order to understand the mTORC2 activation. Modeling enabled hypothesizing a new mechanism, which could reproduce all the findings. The new mechanism was then validated in subsequent experiments [39].

The GerontoMitoSys project tracks the eluding principles of aging by examining the consequences of protein quality and ROS in fungus, mouse, and man (F in fig. 3). Advanced statistical techniques were employed on transcriptome data of the fungus *Podospora anserina* in an attempt to identify global expression patterns. The data analysis indicated that the mTOR signaling pathway turned out to be important during aging, which provided protein quality control through autophagy [40].

ROS – The Dose Makes the Poison

The GerontoMitoSys project finds that in addition to autophagy, protein quality control is also important to delay detrimental effects of aging [40]. Among the principal agents of protein damage are ROS. In order to further elucidate the role of ROS in the emergence of aging-associated phenotypes, a mathematical model was generated using a set of differential equations describing the time course of ROS, its scavengers like cellular antioxidants (SOD) and a protease involved in protein quality control (F2 in fig. 3). Model simulations indicated that the mechanism by which ROS mediates cellular damage is far more complicated than originally stated by the free radical theory of aging. In the paper, the authors suggest that a systems biology approach, similar to the one deployed in the paper, is the one required for an effective integration of the various pathways known to be involved in the control of biological aging.

Although higher ROS levels contribute to accumulation of damage with age, ROS at lower levels trigger defense mechanisms and even provide life extension [41]. These dose-dependent antagonistic actions of ROS are particularly interesting from a systems biology perspective, though explicit systemic studies have yet to be published. In line with this, two separate endogenous mechanisms of ROS production in *Caenorhabditis elegans* were recently described by the JenAge initiative (G). One mechanism commences with impaired insulin and IGF-1 signaling [42], the other is rooted in the deacetylation reaction catalyzed by the sirtuin Sir-2.1 [43].

Taken together, we have shown that several projects within the Gerontosys initiative have proved to make significant contributions in aging research applying systems biology approaches that rely on either advanced analysis of high-throughput data or mathematical modelling of aging-relevant signaling pathways. We think that systems biology in aging is particularly challenging because of the multiple aging phenotypes that exist, the large and often poorly characterized regulatory networks involved and the multiple levels of organization cross talking in the emergence of aging. To make the situation more cumbersome, even for aging processes where the involved pathways are fairly well characterized, experimentally measured output may lead to ambiguous results, because of unexpected interferences by other pathways. It is therefore challenging to devise and plan interdisciplinary research projects in the field of aging. To substantiate this claim, in the next section we discuss the concept, development, and challenges of implementing systems biological approaches in a real case study, a project dealing with the interplay between oxidative phosphorylation, ROS production and aging effects in mice that we have developed over the last 4 years.

Design Concept, Development and Challenges of a Systems Biology Aging Project: The Case Study of the ROSage Project

Hypothesis, Goals and Strategy

The ROSage project has been also part of the Gerontosys Initiative. The acronym of the project stands for 'Reactive Oxygen Species and the Dynamics of Ageing'. The underlying hypothesis of the project, similar to others already mentioned, was that mitochondrial dysfunction and subsequent ROS production are crucial players in the emergence of aging phenotypes via the induction of DNA damage, inflammation and, ultimately, cell senescence and tissue dysfunction, in a probably tissue-dependent manner. The project put emphasis on the role of mutations of the proteins integrated in the mitochondrial respiratory chain complexes as drivers of age-associated dysfunction. Published results indicate that each complex causes diverse but still unique

Table 1. Proteins and protein complexes in the focus of the ROSage research initiative. ROSage used defined mutations of components of the oxidative phosphorylation with phenotypes of clinical significance

Complex	Mutations and clinical implications
Complex I (NADH:ubiquinone oxidoreductase)	complex I mutations are responsible for about 40% of mitochondrion-associated clinical pathologies; however, effects of complex I mutations are diffuse; whereas Koopman's group observe a fragmentation of mitochondria and ROS increase in fibroblast cells, others fail to observe these effects, and only find reduced oxygen consumption and delayed recovery following membrane uncoupling stress
Complex III (coenzyme Q:cytochrome c oxidoreductase)	this complex is required as an oxygen sensor that produces ROS upon hypoxia, and its dysfunction can cause exercise intolerance in humans
Complex IV (cytochrome C oxidase)	a polymorphism with a substitution of nt9348A (V>I) increases the ATP level in the cells which is associated with protection against colitis and A-β accumulation during Alzheimer disease progression
Complex V (ATP synthase)	the substitution of nt778T (D>Y) of mice is associated with multiple aging-associated autoimmune diseases like diabetes, nephritis, pancreatitis, arthritis
Uncoupling protein (UCP2)	strains with dysfunctional UCP2 easily develop autoimmune and chronic inflammatory diseases, while its upregulation confers long life

phenotypes in an organ-specific manner, but also different organs seem to have different energy requirements and ROS sensitivities (see table 1).

To further substantiate our hypothesis, we conceived a complex systems biology workflow, which combines systematic measurements of intracellular signals and tissue markers of organ function during the life span of mice, advanced data analysis and mathematical modelling of selected signaling pathways.

As experimental system, we used several conplastic mouse strains, which hold defined stable mtDNA mutations in several critical proteins involved in the respiratory chain complexes [44]. Since the mitochondrial respiratory chain complexes carry out oxidative phosphorylation, and its malfunctioning can generate excessive ROS, we expected to generate valuable data on the relation between the overproduction of ROS, the activation of cell damage signaling pathways and the triggering of cell senescence and certain tissue-specific aging phenotypes. To this end and for different organs whose dysfunction is associated with aging, we have measured over the life span of the mice: (a) molecular and functional markers of the mitochondrial function, including ROS production, oxygen consumption, quantification of OXPHOS components, intracellular calcium dynamics, mitochondrial dynamics; (b) ROS-mediated activation of signaling pathways involved in damage response like p53, NFκB and JNK; (c) markers for the triggering of phenotypes of cell cycle arrest, senescence and apoptosis, and (d) tissue-specific markers of tissue function, for example neuronal plasticity and learning behavior to account for the phenotypic effects of mitochondria dysfunction in the brain (fig. 4). Five model systems of organ aging were considered: brain function (neurode-

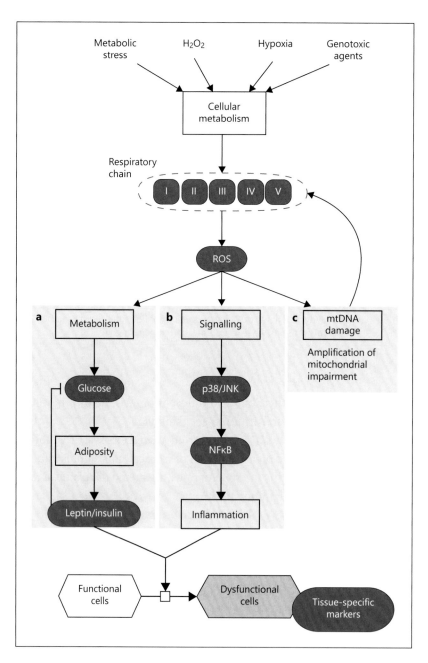

Fig. 4. Regulatory map of the modelling strategy proposed for the ROSage project. The model investigates how enhanced ROS production due to impairment of respiratory chain complexes (I, II, III, IV, V) affects energy homeostasis (**a**), activates inflammatory responses leading to cellular dysfunction and senescence (**b**), and induces further impairment of mitochondrial function (**c**). The interplay between these molecular mechanisms inducing metabolism and inflammation is being further characterized and refined in modules **a–c**.

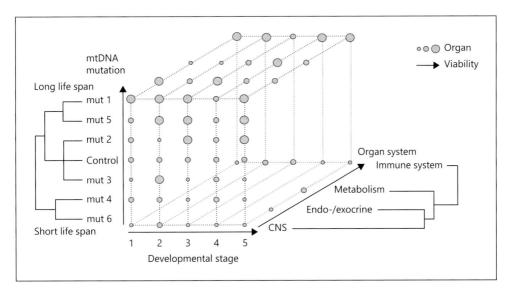

Fig. 5. Systematic measurement of experimental parameters across the dimensions of mitochondrial mutations, organ and developmental stage (age) proposed for the Gerontosys project ROSage. Each dot in the grid represents a set of biomedical insights; the size of the dot corresponds to organ function/viability measured independently for the same mutation, organ and age. Given a similarity measure, a clustering based on modeling insights yields further understanding.

generation), liver- and pancreatic function (metabolic syndrome), pancreatic function (inflammation), hematopoietic system (autoimmune processes) and skin.

To analyze these data, we designed a systems biology workflow that combined advanced data analysis techniques and mathematical modelling. Concerning data analysis, we have been using unsupervised clustering and machine learning techniques to dissect the three-dimensional data grid, which emerges when measuring the indicated markers over the life span for the different organs and developmental stages in the mice strains with different mtDNA mutations (fig. 5). The aim is to find systemic patterns relating ROS generation across time, organ and mitochondrial gene mutations, as well as their relation to organ function/viability. This analysis was complemented with the derivation and characterization of mathematical models accounting for critical parts of the ROS-related network proposed.

Precisely, the network was organized into interconnected modules, accounting for mitochondrial performance, metabolism and inflammation (fig. 4a–c, respectively).

An Overall Perspective on the Current Biological Insights Generated in the Project

The project is still in progress, and most of the data related to life span-related changes in the mice strains are still to be processed. Nevertheless, preliminary results confer to mutations in complexes I, IV, and V, known to be important contributors to a healthy physiology, clear tissue-dependent effects; for example, only the central neural system (CNS) differences between control mouse strains and those mice carrying a mutation in complex I (NADH:ubiquinone oxidoreductase). The differences are a higher ROS level in early life, which reduced to normal level from middle age onwards. Also during middle age, mutant mouse strains displayed reduced plasticity and learning performance.

Mutations in complex IV (cytochrome C oxidase) reduced learning performance in the CNS as well, but the effects were only apparent late in life (24 months). However, no differences in ROS levels were detectable. In addition to the central nervous system, complex IV profoundly affected the liver. By the age of 6 months, mitochondria start to aggregate and ROS as well as antioxidants increase.

Mutations in complex V (ATP synthase) affected the hematopoietic system and the endocrine pancreas. The ROS level was increased in the β-cells of the endocrine pancreas, whereas the ATP level was decreased, although the ATP/ADP ratio was comparable to the wild type [45]. In a high-fat diet condition, the mitochondria grow in size in β-cells of the wild type, but not in complex V mutants. The mutants were also subjected to reductions in insulin secretion and insulin sensitivity along with higher serum insulin concentrations. To counterbalance the loss in insulin secretion due to the lower absolute ATP level, complex V mutants experienced an increase in the total β-cell mass [45]. In the hematopoietic systems of aged mice, the effect of mutations in complex V was surprisingly different – ROS was reduced and ATP increased. ROS reduction and ATP increase could be the result of a metabolic switch to energy conservation in a background with ATP synthase mutations.

Taken together, our preliminary results indicate that (a) overall, organs and tissues are surprisingly robust regarding the mtDNA mutations investigated, and (b) the effect of mutations is tissue-dependent. The phenotypes of mutants are often comparable to the wild type in conventional rearing conditions over most of the mouse life span. This resilience to genetic perturbation highlights the extraordinary capacity of the OXPHOS system and the metabolism in general to balance energy demands. In particular, the mutations of complex III and UCP2 are buffered well because for conventional treatment, physiological conditions and ROS level were comparable in the four examined organs for mutant and wild-type strains.

Results Derived from the Modeling Approach Used

As mentioned before, in the project, the network was organized into interconnected modules, accounting for mitochondrial performance, metabolism and inflammation (fig. 4). We have derived mathematical models for these modules and used them to investigate critical features of the modules associated with aging. For example, combining modeling and experiments, we demonstrated that miRNAs, whose deregulation has been related to aging [46], can modulate, alone or synergistically, the activa-

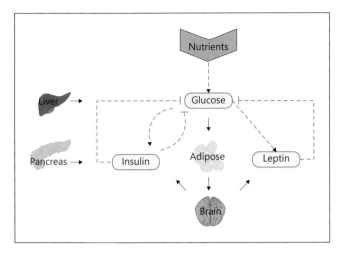

Fig. 6. Multiple levels of structural and functional organization of nutrient-induced glucose metabolism during aging. Glucose is produced (mainly by liver), distributed and utilized by pancreas-induced insulin and adipose tissue-induced insulin and leptin. High-fat nutrient results in higher energy intake, which increases fat accumulation in adipose tissues. Increased fat mass is sensed by the brain, which then signals an increase in leptin and insulin levels in blood for a more rigorous utilization of glucose by other tissues so as to bring glucose levels to levels prior to perturbation. With our aging mice strains, we are deriving and characterizing a mathematical model of this system, which describes the role played by systemic signals like leptin and insulin in modulating cross talk between the organs involved in glucose metabolism and the overall regulation of the glucose metabolism. Interestingly, the system is enriched in insulin- and leptin-mediated feedback loops, which we think control the nonlinear behavior in the blood glucose levels of several strains of mice used in our project (negative feedback loops: glucose → insulin ⊣ glucose; glucose → leptin ⊣ glucose).

tion of p21 by signals activating the DNA damage response like ROS [9, 47]. Our model simulations indicate that miRNAs can posttranscriptionally regulate p21 basal levels in a biological context-dependent manner. Since p21 has been recently involved together with ROS in a feedback loop system able to trigger cell senescence upon a sufficiently long-lasting DNA damage response [21], one can expect miRNA regulation to play a role in this process.

Although this is a promising result, most of the modeling effort in the project has focused on the relation between ROS and metabolic performance (fig. 4b). Despite glucose homeostasis being so crucial for long-term metabolic consequences, there are only a few models which deal with the long-term changes during the life span [48]. The paucity of long-term models is due to several challenges. One of the formidable challenges is that glucose homeostasis involves multiple levels of structural and functional organization, from molecular reactions and cell-cell interactions in tissues like pancreas and liver to the physiology of the entire organs. Figure 6 depicts the involvement of several tissues and organs in carrying out the glucose insulin metabolism in response to nutrient stress, during the life span of a mouse. Although disperse, unconnected mathematical models exist in the literature at all levels of organization [49]; the real challenge here is to integrate multiple levels of data, models and knowledge into a comprehensive multi-level model able to interpret data on the evolution of glucose homeostasis over life span. Multi-levelness and integration of knowledge can be addressed through multi-scale modeling, a topic strongly developed in the last years in cancer biology, but still in its infancy for most of the other biomedical fields, including aging research.

In line with the idea and in the context of the ROSage project, we have been analyzing the relation between glucose metabolism and mitochondria performance in aging by deploying a cascade of models describing different levels of organization. At the physiological level (fig. 6), we have developed a phenomenological mesoscopic mathematical model, which accounts for the interplay between different critical organs (liver, pancreas, adipose tissue, brain) via molecules that act in this context as systemic signals, secreted by these organs and used to mediate their cross talk and the overall regulation of the glucose metabolism (insulin, leptin). The mathematical modelling has been used to establish the role played by several insulin- and leptin-mediated feedback loops in shaping nonlinear behavior in the glucose blood levels of several of our aging mouse strains.

In parallel to this, we built upon one of the assumptions of damage accumulation in the mtDNA during the life span and their effect on the mitochondrial performance through its effect in the balance between the antagonizing forces of mitochondrial fission and fusion [50] Mitochondria organize themselves as dynamic populations within a cell, by undergoing continuous cycles of fission and fusion. In several neurodegenerative and metabolic diseases, the dynamic balance of fission and fusion is disturbed. A diversified mitochondrial population is generated by the heterogeneous availability of nutrients and signals in cellular cross-section [51]. Segregation through fission increases local sensitivity of mitochondria towards signals, whereas fusion interconnects mitochondrial networks from one location to another, hence increasing their global sensitivity. One can say that global coupling of excitatory and restoring mitochondrial subpopulations in a heterogeneous environment enables the system to selectively adapt the response at one location, while amplifying the response at another location.

We were interested in establishing how the equilibrium between fission and fusion is altered over the life span of the investigated mouse strains. To this end, we established a mathematical model of antagonistic fission and fusion subpopulations of mitochondria by adapting the model proposed by Wallach and coworkers to investigate spatially heterogeneous neuronal systems [52].

Fission and fusion subpopulations encounter heterogeneous signals in the peripheral (high frequency signal) and perinuclear (low frequency signal) cellular locations [51]. In our mathematical model, we divide the mitochondrial subpopulations into three: peripheral fission, perinuclear fission and global fusion subpopulations (see fig. 7a for graphical depiction of the model and explanation of the mathematical equations). Two fission subpopulations interact antagonistically with the global fusion subpopulation (fig. 7a). Model simulations and analysis indicate that global coupling of mitochondrial subpopulations in a heterogeneous environment enables the system to

Chauhan · Liebal · Vera · Baltrusch · Junghanß · Tiedge · Fuellen · Wolkenhauer · Köhling

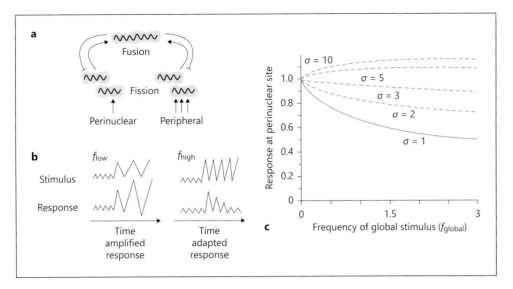

Fig. 7. a Our model of mitochondria performance includes two antagonistic subpopulations of dividing (fission) and fusing mitochondria, which are globally coupled through fusion in the heterogeneous cellular environment. At the peripheral locations, high-frequency stimuli are prevalent, while at the perinuclear location low-frequency stimuli are prevalent. In our mathematical model, we divide the mitochondrial subpopulations into three: peripheral fission, perinuclear fission and global fusion subpopulations. Two fission subpopulations interact antagonistically with global fusion subpopulation. The three subpopulations are represented by three equations of following type:

$$(1)\ \frac{dP}{dt} = \underbrace{\frac{1-P}{\tau}}_{\text{recovery}} - \underbrace{P \cdot \psi(P, f)}_{\text{depletion}}.$$

P denotes the availability of respective subpopulation: peripheral fission, perinuclear fission or global fusion. The availability of a subpopulation depends on the time an average mitochondrial subpopulation takes to recover, and on the depletion of resources in maintaining the activity *(Ψ)* of respective average subpopulation. Activity is a function of availability of subpopulations *(P)* and external stimulation *(f)*, such that fission subpopulations are antagonistic to the global fusion subpopulation. τ represents the recovery time constant. Next, mitochondrial response is defined as the ratio of activity upon stimulation with a certain frequency *(f)* and activity without stimulation (0).

$$(2)\ \text{Response} = \frac{\psi(P, f)}{\psi(P, 0)}.$$

In the model, we introduce a parameter δ to describe the degree of variability between the peripheral (f_{high}) and perinuclear (f_{low}). δ is defined by a ratio of global frequency f_{global} ($f_{high} + f_{low}$) and f_{low}. f_{global} is the frequency with which fusion population is stimulated as a result of global coupling of stimuli:

$$(3)\ \delta = \frac{f_{global}}{f_{low}}.$$

b Schematic representation of two types of stimuli (f_{low}, f_{high}) and their responses. **c** Model simulations. Mitochondrial Ca^{2+} levels adapt (**b** and $\delta = 1–3$ in **c**) to high-frequency stimuli, whereas in response to low-frequency stimulation mitochondrial Ca^{2+} levels amplify (**b** and $\delta = 5–10$ in **c**).

selectively adapt the response at one location while amplifying the response at another location (fig. 7b). In figure 7c, a model simulation is shown. Here, we can see that adaptation emerges in the response of the system to low scenarios of variability between the frequency of peripheral and perinuclear stimuli driving fission-fusion cycles ($\delta = 1$–3). In contrast, amplification appears for scenarios of high variability ($\delta = 5$–10).

At the moment, we are elaborating a strategy to design experiments able to test our model predictions. Furthermore, we are expanding our model to account for the connection between the dynamics of mitochondrial subpopulations and the response of critical cell/tissues to environmental perturbations on the longer timescales that correspond to the emergence of aging phenotypes.

Conclusions

In this chapter, we support the idea that boosting aging research will require the use of systems biology-inspired approaches because (a) the dysfunction of large, interconnected biochemical networks is in the origin of aging-associated phenotypes; (b) these networks are enriched in nonlinear regulatory motifs like positive and negative feedback loops; (c) aging manifests at multiple, interconnected and interdependent levels of organization in the body, from the intracellular machinery to the dynamics of tissue organization and beyond, and (d) the optimal design of biomedical strategies to counteract aging-associated pathologies will require the use of tools and strategies adapted from engineering.

The specific problems that one will face when designing a systems biology project in aging are not minor. The difficulty to generate reliable experimental data for the investigation of aging phenotypes increases when thinking from a systemic view because one will require more systematic and frequent and better quantifiable experimental measurements, but also an increase in the number of experimental replicates. The strategy to overcome the difficulties of integrating in a coherent manner quantitative data produced at multiple levels of organization to measure the emergence of aging phenotypes is still an open question. In line with this, the strategy to adapt the multi-scale methodologies used in other biomedical fields like cancer is still to be established. The question of establishing reliable methodologies for actual quantification of aging phenotypes at the intracellular, tissue and organ levels is open to debate. Finally, large and long-lasting collaborative systems biology projects like those addressing the molecular basis of aging can only prosper when a good data management strategy is designed prior to project initiation.

Acknowledgements

This work was supported by the German Federal Ministry of Education and Research as part of the project Gerontosys-ROSage (0315892A).

Chauhan · Liebal · Vera · Baltrusch · Junghanß · Tiedge · Fuellen · Wolkenhauer · Köhling

References

1 Vera J, Wolkenhauer O: A system biology approach to understand functional activity of cell communication systems. Methods Cell Biol 2008;90:399–415.

2 Vera J, Rath O, Balsa-Canto E, Banga JR, Kolch W, Wolkenhauer O: Investigating dynamics of inhibitory and feedback loops in ERK signalling using power-law models. Mol Biosyst 2010;6:2174–2191.

3 Schoeberl B, Pace EA, Fitzgerald JB, Harms BD, Xu L, Nie L, Linggi B, Kalra A, Paragas V, Bukhalid R, Grantcharova V, Kohli N, West KA, Leszczyniecka M, Feldhaus MJ, Kudla AJ, Nielsen UB: Therapeutically targeting ErbB3: a key node in ligand-induced activation of the ErbB receptor-PI3K axis. Sci Signal 2009;2:ra31.

4 Vera J, Schmitz U, Lai X, Engelmann D, Khan FM, Wolkenhauer O, Pützer BM: Kinetic modeling-based detection of genetic signatures that provide chemoresistance via the E2F1-p73/DNp73-miR-205 network. Cancer Res 2013;73:3511–3524.

5 Ptitsyn AA, Weil MM, Thamm DH: Systems biology approach to identification of biomarkers for metastatic progression in cancer. BMC Bioinformatics 2008;9(suppl 9):S8.

6 Sara H, Kallioniemi O, Nees M: A decade of cancer gene profiling: from molecular portraits to molecular function. Methods Mol Biol 2010;576:61–87.

7 Vera J, Lai X, Schmitz U, Wolkenhauer O: MicroRNA-regulated networks: the perfect storm for classical molecular biology, the ideal scenario for systems biology. Adv Exp Med Biol 2013;774:55–76.

8 Meyer P, Cokelaer T, Chandran D, Kim KH, Loh P-R, Tucker G, Lipson M, Berger B, Kreutz C, Raue A, Steiert B, Timmer J, Bilal E, Sauro HM, Stolovitzky G, Saez-Rodriguez J: Network topology and parameter estimation: from experimental design methods to gene regulatory network kinetics using a community based approach. BMC Systems Biol 2014;8:13.

9 Lai X, Schmitz U, Gupta SK, Bhattacharya A, Kunz M, Wolkenhauer O, Vera J: Computational analysis of target hub gene repression regulated by multiple and cooperative miRNAs. Nucleic Acids Res 2012; 40:8818–8834.

10 Tyson JJ, Chen KC, Novak B: Sniffers, buzzers, toggles and blinkers: dynamics of regulatory and signaling pathways in the cell. Curr Opin Cell Biol 2003;15: 221–231.

11 Nikolov S, Wolkenhauer O, Vera J: Tumors as chaotic attractors. Mol Biosyst 2014;10:172–179.

12 Aldridge BB, Burke JM, Lauffenburger DA, Sorger PK: Physicochemical modelling of cell signalling pathways. Nat Cell Biol 2006;8:1195–1203.

13 Kreeger PK, Lauffenburger DA: Cancer systems biology: a network modeling perspective. Carcinogenesis 2010;31:2–8.

14 Byrne HM: Dissecting cancer through mathematics: from the cell to the animal model. Nat Rev Cancer 2010;10:221–230.

15 Engel C, Scholz M, Loeffler M: A computational model of human granulopoiesis to simulate the hematotoxic effects of multicycle polychemotherapy. Blood 2004;104:2323–2331.

16 Kirkwood TB, Kowald A: Network theory of aging. Exp Gerontol 1997;32:395–399.

17 Kowald A, Kirkwood TB: Towards a network theory of ageing: a model combining the free radical theory and the protein error theory. J Theor Biol 1994;168: 75–94.

18 Franceschi C, Bonafè M, Valensin S, Olivieri F, De Luca M, Ottaviani E, De Benedictis G: Inflamm-aging. An evolutionary perspective on immunosenescence. Ann N Y Acad Sci 2000;908:244–254.

19 de Magalhães JP, Toussaint O: GenAge: a genomic and proteomic network map of human ageing. FEBS Lett 2004;571:243–247.

20 Xue H, Xian B, Dong D, Xia K, Zhu S, Zhang Z, Hou L, Zhang Q, Zhang Y, Han J-DJ: A modular network model of aging. Mol Syst Biol 2007;3:147.

21 Passos JF, Nelson G, Wang C, Richter T, Simillion C, Proctor CJ, Miwa S, Olijslagers S, Hallinan J, Wipat A, Saretzki G, Rudolph KL, Kirkwood TB, von Zglinicki T: Feedback between p21 and reactive oxygen production is necessary for cell senescence. Mol Syst Biol 2010;6:347.

22 Kriete A, Bosl WJ, Booker G: Rule-based cell systems model of aging using feedback loop motifs mediated by stress responses. PLoS Comput Biol 2010;6: e1000820.

23 Kitano H: Biological robustness. Nat Rev Genet 2004;5:826–837.

24 Kirkwood TB: Understanding the odd science of aging. Cell 2005;120:437–447.

25 Kirkwood TB: A systematic look at an old problem. Nature 2008;451:644–647.

26 Passarino G, Rose G, Bellizzi D: Mitochondrial function, mitochondrial DNA and ageing: a reappraisal.' Biogerontology 2010;11:575–588.

27 Hoehme S, Brulport M, Bauer A, Bedawy E, Schormann W, Hermes M, Puppe V, Gebhardt R, Zellmer S, Schwarz M, Bockamp E, Timmel T, Hengstler JG, Drasdo D: Prediction and validation of cell alignment along microvessels as order principle to restore tissue architecture in liver regeneration. Proc Natl Acad Sci USA 2010;107:10371–10376.

28 van Leeuwen IM, Vera J, Wolkenhauer O: Dynamic energy budget approaches for modelling organismal ageing. Philos Trans R Soc Lond B Biol Sci 2010;365: 3443–3454.

29 van Leeuwen IM, Kelpin FD, Kooijman SA: A mathematical model that accounts for the effects of caloric restriction on body weight and longevity. Biogerontology 2002;3:373–381.

30 McAuley MT, Wilkinson DJ, Jones JJ, Kirkwood TB: A whole-body mathematical model of cholesterol metabolism and its age-associated dysregulation. BMC Syst Biol 2012;6:130.

31 De Grey AD, Rae M: Ending Aging: The Rejuvenation Breakthroughs That Could Reverse Human Aging in Our Lifetime. New York, St. Martin's Griffin, 2008.

32 De Grey AD, Ames BN, Andersen JK, Bartke A, Campisi J, Heward CB, McCarter RJM, Stock G: Time to talk SENS: critiquing the immutability of human aging. Ann N Y Acad Sci 2002;959:452–462.

33 Wang J, Sun Q, Morita Y, Jiang H, Gross A, Lechel A, Hildner K, Guachalla LM, Gompf A, Hartmann D, Schambach A, Wuestefeld T, Dauch D, Schrezenmeier H, Hofmann W-K, Nakauchi H, Ju Z, Kestler HA, Zender L, Rudolph KL: A differentiation checkpoint limits hematopoietic stem cell self-renewal in response to DNA damage. Cell 2012;148:1001–1014.

34 Przybilla J, Galle J, Rohlf T: Is adult stem cell aging driven by conflicting modes of chromatin remodeling? Bioessays 2012;34:841–848.

35 Przybilla J, Rohlf T, Loeffler M, Galle J: Understanding epigenetic changes in aging stem cells – a computational model approach. Aging Cell 2014;13:320–328.

36 Dodd IB, Micheelsen MA, Sneppen K, Thon G: Theoretical analysis of epigenetic cell memory by nucleosome modification. Cell 2007;129:813–822.

37 Cornu M, Albert V, Hall MN: mTOR in aging, metabolism, and cancer. Curr Opin Genet Dev 2013;23:53–62.

38 Grahammer F, Wanner N, Huber TB: mTOR controls kidney epithelia in health and disease. Nephrol Dial Transplant 2014;29(suppl 1):i9–i18.

39 Dalle Pezze P, Sonntag AG, Thien A, Prentzell MT, Gödel M, Fischer S, Neumann-Haefelin E, Huber TB, Baumeister R, Shanley DP, Thedieck K: A dynamic network model of mTOR signaling reveals TSC-independent mTORC2 regulation. Sci Signal 2012;5:ra25.

40 Philipp O, Hamann A, Servos J, Werner A, Koch I, Osiewacz HD: A genome-wide longitudinal transcriptome analysis of the aging model podospora anserine. PLoS One 2013;8:e83109.

41 Wink DA, Hines HB, Cheng RYS, Switzer CH, Flores-Santana W, Vitek MP, Ridnour LA, Colton CA: Nitric oxide and redox mechanisms in the immune response. J Leukoc Biol 2011;89:873–891.

42 Zarse K, Schmeisser S, Groth M, Priebe S, Beuster G, Kuhlow D, Guthke R, Platzer M, Kahn CR, Ristow M: Impaired insulin/IGF1 signaling extends life span by promoting mitochondrial L-proline catabolism to induce a transient ROS signal. Cell Metab 2012;15:451–465.

43 Schmeisser K, Mansfeld J, Kuhlow D, Weimer S, Priebe S, Heiland I, Birringer M, Groth M, Segref A, Kanfi Y, Price NL, Schmeisser S, Schuster S, Pfeiffer AFH, Guthke R, Platzer M, Hoppe T, Cohen HY, Zarse K, Sinclair DA, Ristow M: Role of sirtuins in lifespan regulation is linked to methylation of nicotinamide. Nat Chem Biol 2013;9:693–700.

44 Yu X, Gimsa U, Wester-Rosenlof L, Kanitz E, Otten W, Kunz M, Ibrahim SM: Dissecting the effects of mtDNA variations on complex traits using mouse conplastic strains. Genome Res 2008;19:159–165.

45 Weiss H, Wester-Rosenloef L, Koch C, Koch F, Baltrusch S, Tiedge M, Ibrahim S: The mitochondrial Atp8 mutation induces mitochondrial ROS generation, secretory dysfunction, and β-cell mass adaptation in conplastic B6-mtFVB mice. Endocrinology 2012;153:4666–4676.

46 Smith-Vikos T, Slack FJ: MicroRNAs and their roles in aging. J Cell Sci 2012;125:7–17.

47 Lai X, Bhattacharya A, Schmitz U, Kunz M, Vera J, Wolkenhauer O: A systems biology approach to study microRNA-mediated gene regulatory networks. Biomed Res Int 2013;703849:1–15.

48 Goldbeter A: A model for the dynamics of human weight cycling. J Biosci 2006;31:129–136.

49 Wolkenhauer O: The role of theory and modeling in medical research. Front Physiol 2013;4:377.

50 Chauhan A, Vera J, Wolkenhauer O: The systems biology of mitochondrial fission and fusion and implications for disease and aging. Biogerontology 2014;15:1–12.

51 Collins TJ, Berridge MJ, Lipp P, Bootman MD: Mitochondria are morphologically and functionally heterogeneous within cells. EMBO J 2002;21:1616–1627.

52 Wallach A, Eytan D, Marom S, Meir R: Selective adaptation in networks of heterogeneous populations: model, simulation, and experiment. PLoS Comput Biol 2008;4:e29.

Dr. Anuradha Chauhan
Deptartment of Systems Biology & Bioinformatics
Institute of Computer Science
University of Rostock
DE-18051 Rostock (Germany)
E-Mail anuradha.chauhan@uni-rostock.de

Yashin AI, Jazwinski SM (eds): Aging and Health – A Systems Biology Perspective.
Interdiscipl Top Gerontol. Basel, Karger, 2015, vol 40, pp 177–188 (DOI: 10.1159/000364982)

Conservative Growth Hormone/IGF-1 and mTOR Signaling Pathways as a Target for Aging and Cancer Prevention: Do We Really Have an Antiaging Drug?

Vladimir N. Anisimov

Department of Carcinogenesis and Oncogerontology, N.N. Petrov Research Institute of Oncology,
St. Petersburg, Russia

Abstract

Inactivation of the GH/insulin/IGF-1 signaling molecules corresponding genes as well as the inactivation of serine/threonine protein kinase mTOR increases life span in nematodes, fruit flies and mice. Evidence has emerged that antidiabetic biguanides and rapamycin are promising candidates for pharmacological interventions leading to both life span extension and prevention of cancer. The available data on the relationship of two fundamental processes – aging and carcinogenesis – have been suggested to be a basis for understanding these two-side effects of biguanides and rapamycin.

© 2015 S. Karger AG, Basel

There are nine tentative hallmarks of aging in mammals, which may represent common denominators of aging in different organisms: genomic instability, telomere attrition, epigenetic alterations, loss of proteostasis, deregulated nutrient sensing, mitochondrial dysfunction, cellular senescence, stem cell exhaustion, and altered cell-to-cell communication [1]. At the same time, there is also sufficient similarity in the patterns of changes observed during normal aging and the process of carcinogenesis (table 1) [2]. As can be seen in figure 1, DNA damage induced by environmental and endogenous carcinogenic factors [reactive oxygen species, ionizing radiation, ultraviolet, constant illumination (light at night), some diets, oncogenes, etc.] may lead to cellular senescence or cellular lesions which could be deleted by apoptosis. The same agents can induce damage which is followed by neoplastic transformation, thus leading to cancer [2, 3]. During the last decade, the intensive search for antiaging remedies has led to the conclusion that both insulin/IGF-1 signaling (IIS) and nutrient response pathways defined by the mTOR protein kinase pathways control aging and age-asso-

Table 1. Changes developing in organism during natural aging and carcinogenesis: effects of gero-protectors

Parameters	Aging	Carcinogenesis	Biguanides	Rapamycin
Molecular level				
Free radical generation	↑	↑	↓	↓
AGE formation	↑	↑	↓	↓
DNA adduct formation	↑	↑	↓	↓
DNA repair efficacy	↓	↓	↓	↓
Genomic instability	↑	↑	↓	↓
Telomerase activity	↓	↑	↓	↓
Telomere length	↑	↓	↑	↑
mTOR activity	↑	↑	↓	↓
IKK-β/NF-κB activity	↑	↑	↓	↓
Clock gene expression (*Per1, Per2*)	↓	↓	↓	↑
Mutation rate	↑	↑	↓	?
Oncogene expression	↑	↑	↓	↓
p53 mutations	↓	↑	?	?
Cellular/tissue level				
Oxidative stress	↑	↑	↓	↓
Chromosome aberrations	↑	↑	↓	↑
Induced pluripotent stem cells	↓	↓	↑	↑
Proliferative activity	↓	↑	↓	↓
Focal hyperplasia	↑	↑	↓	↓
Apoptosis	↓	↓	↑	↑
Autophagy	↓	↓	↑	↑
Angiogenesis	↓	↓	↓	↓
Cell-to-cell communication	↓	↓	↑	↑
Senescent cells number	↑	↑	↓	↓
Latent (dormant) tumor cell number	↑	↑	↓	↓
Systemic/organism level				
Melatonin circadian rhythm	↓	↓	↓	?
Serum melatonin level	↓	↓	↓	?
Hypothalamic threshold of sensitivity to homeostatic inhibition by steroids	↑	↑	↓	?
Tolerance to glucose	↓	↓	↑	↓
Serum insulin level	↑	↑	↓	↓
Susceptibility to insulin	↓	↓	↑	↑
LDL and cholesterol level	↑	↑	↓	↓
Ovulatory function	↓	↓	↑	↑
Fertility	↓	↓	↑	?
T cell immunity	↓	↓	↑	↓
Inflammation	↑	↑	↓	↓
Cancer risk	↑	↑	↓	↓
Life span	↓	↓	↑	↑

↑ = Increases; ↓ = decreases; ? = no data.

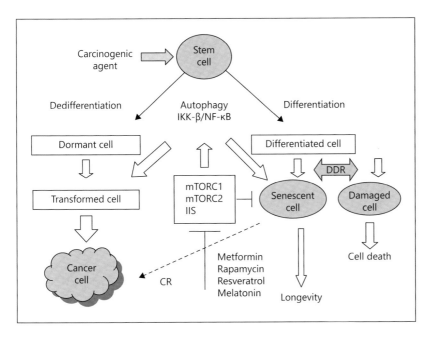

Fig. 1. Relationship between aging and carcinogenesis: the key role of IIS and mTOR signaling. DNA damage induced by environmental and endogenous factors (reactive oxygen species, ionizing radiation, ultraviolet, constant illumination, some diets, oncogenes, etc.) may lead to cellular senescence or cellular lesions which could be deleted by apoptosis. The same agents can induce damage followed by neoplastic transformation, thus leading to cancer. Metformin, rapamycin, and some other compounds with mTOR and IIS-inhibitory potential (resveratrol, melatonin) are able to modify both aging and carcinogenesis. DDR = DNA damage response.

ciated pathology in worms, insects and mammals [4]. In each of these organisms, genetic downregulation or interruption of this signaling pathway can lead to major extension of longevity. There are two functionally distinct mTOR complexes called mTORC1 and mTORC2. mTORC1 is activated by insulin and related growth factors through phosphatidylinositol-3-OH kinase and AKT kinase signaling and repressed by AMP-activated protein kinase, a key sensor of cellular energy status [4]. The mTORC1 is involved in promoting messenger RNA translation and protein synthesis through ribosomal protein S6 kinases (S6Ks) and 4E-BP protein, which in the hypophosphorylated form acts as a negative regulator of the cap-binding protein eIF4E. mTORC1 also stimulates lipid biosynthesis, inhibits autophagy, and through hypoxic response transcription factor HIF-1α regulates mitochondrial function and glucose metabolism. Rapamycin suppresses mTORC1 and indirectly mTORC2, which leads to metabolic lesions like glucose intolerance and abnormal lipid profile [4]. The phosphorylation of S6K1 at T389 by TORC1 is susceptible to rapamycin. The life span of S6K1-deficient female mice increased by 19% in comparison to the wild-type controls without any effect on the incidence of tumor development [5]. It is worth noting that there was no significant effect of the protein knockout on the life span of male mice.

These data suggest that S6K1 is involved in mammalian life span regulation downstream of TORC1. Taking into consideration the negative effect of rapamycin on glucose tolerance and liver insulin sensitivity [5], Lamming et al. [6] studied the effect of mTORC1 and mTORC2 regulator gene modification on the life span of mice. There was no increase in life span in either female or male *mtor*[+/−], *Raptor*[+/−], *mlst8*[+/−] or *mtor*[+/−] *Raptor*[+/−] mice. However, female *mtor*[+/−] *mlst8*[+/−] mice lived longer by 14.4% in comparison to wild-type mice. The longevity of male *mtor*[+/−] *mlst8*[+/−] mice was unaffected. Female *mtor*[+/−] *mlst8*[+/−] mice were not calorie restricted through reduced food intake or increased energy expenditure, and had normal body weights and levels of activity consistent with the phenotypic effects. *mtor*[+/−] *mlst8*[+/−] mice exhibited an approximately 30–60% reduction in the abundance of hepatic mTOR, Raptor, mLST8, and Rictor, whereas the expression of mTOR complex subunits was less affected in *Raptor*[+/−] and *mtor*[+/−] *Raptor*[+/−] heterozygotes. The authors believe that suppression mTORC1 signaling is sufficient for life span prolongation independent of changes in glucose homeostasis.

Calorie restriction (CR) is the only known intervention in mammals that has been consistently shown to increase life span, reduce incidence and retard the onset of age-related diseases, including cancer and diabetes. CR has also been shown to increase resistance to stress and toxicity, and maintain youthful levels of function and vitality in laboratory mammals at advanced chronological age [7]. CR in rhesus monkeys have produced physiological responses strikingly similar to those observed in rodents and delayed the onset of age-related diseases, but effects on longevity were not consistent [8]. Data from these studies indicate that long-term CR reduces morbidity and mortality in primates, and thus may exert beneficial 'antiaging' effects in humans. Although understanding the role of GH and IIS in the control of human aging is incomplete and somewhat controversial, available data indicate that dietary prevention of excessive IGF-1 and insulin secretion and using diet and exercise to enhance insulin sensitivity may represent the most hopeful approaches to cancer prevention and to extending human health span and life span [2–4]. Metformin, rapamycin, and some other compounds with mTOR- and IIS-inhibitory potential (resveratrol, melatonin) are able to modify both aging and carcinogenesis. This chapter focuses on the effects of biguanides and rapamycin, whereas data on resveratrol and melatonin will be discussed elsewhere.

Antidiabetic Biguanides as Geroprotectors

The concept of CR mimetics is now being intensively explored [3]. The antidiabetic biguanides, phenformin, buformin and metformin were observed to reduce hyperglycemia, improve glucose utilization, reduce free fatty acid utilization, gluconeogenesis, serum lipids, insulin, and IGF-1, reduce body weight and decrease metabolic immunodepression both in humans and rodents [3]. The results of studies on

Table 2. Summary on effects of biguanides and rapamycin on life span and spontaneous carcinogenesis in rodents

Strain, species	Sex	C/T mice	Age at start of treatment, months	Drug	Dose and route of treatment	Effect on mean life span, %	Effect on carcinogenesis	Ref.
Biguanides								
C3H/Sn mice	F	30/24	3.5	PF	2 mg/mouse, p.o.	+21	↓	[9]
HER-2/neu mice	F	34/32	2	MF	100 mg/kg, p.o.	+8	↓	[10]
HER-2/neu mice	F	31/35	2	MF	"	+4	↓	[11]
SHR mice	F	50/50	3	MF	"	+38	=	[12]
SHR mice	F	119/51	3	MF	"	+14	=	[13]
		97/45	9	MF	"	+6	=	
		69/33	15	MF	"	0	=	
129/Sv mice	M	41/46	3	MF	"	−13	=	[14]
129/Sv mice	F	47/41	3	MF	"	+5	↓	
C57BL/6	M	64/83	12	MF	0.1% in diet	+5.83	=	[17]
	M	90/88	12	MF	1% in diet	−14.4	↓	
B6C3F1	M	297/36	12	MF	0.1% in diet	+5.83	=	
LIO rats	F	41/44	3.5	PF	5 mg/rat, p.o.	0	↓	[9]
LIO rats	F	74/42	3.5	BF	"	+7	↓	
F344 rats	M	31/40	6	MF	300 mg/kg, food	0	ND	[16]
Rapamycin								
UM-HET3 mice	M	357/134	20	Rap	14 mg/kg, food	+9	=	[18]
UM-HET3 mice	F	289/144	20	Rap	"	+14	=	
UM-HET3 mice	M	50/50	9	Rap	"	+10	=	[19]
UM-HET3 mice	F	50/50	9	Rap	"	+18	=	
HER-2/9eu mice	F	28/30	2	Rap	1.5 mg/kg, s.c.	+4	↓	[20]
129/Sv mice	F	31/35	2	Rap	"	+4	↓	[21]
p53+/− mice	M	38/37	>5	Rap	1.5 mg/kg, d.w.	+10	↓	[22]
p53−/− mice	M	17/21	2	RapT	0.5 mg/kg, p.o.	+30	↓	[23]
Rb1+/− mice	M	97/98	2–3	Rap	14 mg/kg, p.o.	+14	↓	[24]
	F		2–3			+9	↓	

C/T = Control/treatment; BF = buformin; MF = metformin; PH = phenformin; Rap = rapamycin; RapT = Rapatar; d.s. = with drinking water; s.c. = subcutaneously; p.o. = gavage; food = with diet; ↑ = increases; ↓ = decreases; = = no effect; ND = not detected.

the effect of antidiabetic biguanides on the life span in mice and rats are summarized in table 2.

The treatment with phenformin prolonged the mean life span of female C3H/Sn mice by 21% (p < 0.05) and the maximum life span by 26% in comparison with the controls [9]. At the time of death of the last mice in the control group, 42% of phenformin-treated mice were alive. The treatment with phenformin failed to influence the mean life span of female LIO rats; however, it increased the maximum life span by 3 months (10%) in comparison with the controls [9]. The treatment with phenformin slightly decreased the body weight of rats and delayed age-related switching off of estrous function. Similar findings were observed in female rats exposed to another biguanide, buformin [9]. Administration of metformin to female transgenic HER-2/ neu mice did not change the body weight or temperature; it slowed down the age-related rise in blood glucose and triglyceride levels, decreased the serum level of cholesterol and β-lipoproteins, delayed the age-related irregularity in estrous cycle, extended the mean life span by 4–8% and the maximum life span by 1 month in comparison with the control animals [10, 11]. It is well known that excess of body weight and obesity leads to development of metabolic syndrome, type 2 diabetes, premature switching off of reproductive function and risk of cancer [2, 9]. The mechanism behind the geroprotective effect of metformin could reside in its ability to lower body weight.

Metformin increased the mean life span of the last 10% of survivors by 20.8% and the maximum life span by 2.8 months (10.3%) in female SHR mice in comparison with control mice [12]. The decreased body temperature and postponed age-related switching off of estrous function were observed in the group of metformin-treated mice. In another set of experiments, female SHR mice were given metformin from the age of 3, 9 or 15 months [13]. Metformin started at the age of 3 months increased the mean life span by 14% and maximum life span by 1 month, whereas the treatment started at the age of 9 months by 6%; metformin started at the age of 15 months did not affect life span.

The treatment with metformin slightly modified the food consumption but failed to influence the dynamics of body weight. Metformin decreased by 13.4% the mean life span of male 129/Sv mice and slightly increased the mean life span of females (by 4.4%). Metformin failed to influence spontaneous tumor incidence in male 129/Sv mice, decreased 3.5-fold the incidence of malignant neoplasms in female mice, while somewhat stimulated formation of benign vascular tumors [14].

Significant prolongation (by 20.1%) of the survival time was observed in male (but not female) transgenic mice with Huntington's disease without affecting fasting blood glucose levels. Increasing the dose of the drug did not improve the survival of mice [15]. In the NIA study [16], 6-month-old male F344 rats were randomized to one of four diets: control, CR, diet supplemented with metformin and standard diet pair fed to metformin. There were no significant differences in the mean life span of the last surviving 10% of each group in the CR, metformin-treated and pair fed rats as compared with control [16]. CR significantly increased life span in the 25th quantile but not the 50th, 75th, or 90th quantile. The groups of rats exposed to metformin or pair feeding were not significantly different from controls at any quantile. The reduced efficacy of CR in this study might provide a partial explanation for the lack of an increase in life span with metformin.

Male C57BL/6 mice were given ad libitum diet with supplementation of 0.1 or 1% of metformin starting from the age of 12 months until natural death [17]. The mean life span of mice given 0.1% metformin in the diet was increased by 5.83% as compared with the relevant control mice. The 1% dose was toxic and reduced the mean life span by 14.4%. Diet supplementation with 0.1% metformin increased life span by

4.15% in male B6C3F1 mice. There were no significant differences in pathologies observed in both strains of mice fed diet with 0.1% metformin. However, diet with 1% metformin reduced the incidence of liver cancers (3.3% in the metformin group vs. 26.5% in control group, $p < 0.001$). Male C57BL/6 mice given metformin had decreased rates of cataracts [17]. The treatment with metformin mimics some of the benefits of CR, such as improved physical performance, improved glucose-tolerance test, increased insulin sensitivity, and reduced low-density lipoprotein and cholesterol levels without a decrease in caloric intake. Metformin also increased AMP-activated protein kinase activity and increased antioxidant protection, resulting in reductions in both oxidative damage accumulation and chronic inflammation [17]. The administration of metformin to mice induced CR-like genomic and metabolic responses which were interpreted as induction of pathways associated with longevity.

Thus, available data showed that antidiabetic drugs can increase survival of rodents in some cases (table 2). This effect varied depending on the strain and species of animals. Female mice were treated in the majority of these studies. Due to gender differences in the effect of metformin [3], the experiments with males of different strains need to be performed to draw a conclusion regarding the geroprotective potential of antidiabetic biguanides. Only single studies were performed with female rats treated with buformin or phenformin and with male rats treated with metformin. Both male and female animals of different strains need to be treated in the same study for a more exact conclusion on the geroprotective potential of antidiabetic biguanides.

Routes of administration and doses of metformin were different in the majority of experiments discussed here. The NIA (National Institute of Aging, NIH USA) team used diet supplementation with 0.1 or 1% metformin in male C57BL/6 mice and a diet with 0.1% metformin in male B6C3F1 mice [17], whereas the PRIO (Petrov Research Institute of Oncology, St. Petersburg, Russia) team administered metformin with drinking water at a dose of 100 mg/kg to female HER-2/neu mice in two independent studies, to outbreed female Swiss-derived SHR mice also in two sets of experiments, and in male and female inbred 129/Sv mice in one study [10–14] (table 2). Calculations show that C57BL/6 mice consuming the diet with 0.1% metformin received the drug at a dose ranging from 75 to 100 mg/kg body weight, and B6C3F1 mice at a dose between 67 and 90 mg/kg. They are practically the same doses given with drinking water in our studies. The biggest dose of metformin given to C57BL/6 mice (1% in diet) reduced their mean life span by 14.4%, and at first seems toxic for the kidney as it induces its enlargement, lumpiness and decoloration [17]. The NIA team started the treatment with metformin at the age of 12 months, whereas the PRIO team started the treatment at the age of 2–3.5 months in the majority of experiments. When metformin was given to female SHR mice staring at the age of 3, 9 or 15 months, an attenuation of the effect on life span with the increase in the age at start was observed [13].

On the whole, the data in the literature and the results of our experiments suggest that antidiabetic biguanides are promising interventions for slowing down aging and life span extension. Additional studies are required to provide more information on the optimal doses, appropriate age intervals, and other conditions under which exposure to metformin could prevent premature aging in humans.

Effects of Rapamycin on Aging and Longevity

In the National Institute on Aging Intervention Testing Program, male and female genetically heterogeneous mice (UM_HET3) aged 600 days were fed with encapsulated rapamycin in diet [18]. On the basis of age at 90% mortality, rapamycin led to an increase in the mean life span by 14% for females and 9% for males. The authors claimed that disease patterns of rapamycin-treated mice did not differ from those of control mice. It was stressed that rapamycin may extend life span by postponing death from cancer, by retarding aging, or both. However, only a small part of mice was really studied pathomorphologically. In the same paper, similar results of the treatment with rapamycin started at the age of 270 days have been reported. In another set of experiments, rapamycin was given with food (14 mg/kg food; 2.24 mg/kg mouse weight per day) to UM_HET3 mice from the age of 9 months [19]. The mean life span was increased by 10% in males and 18% in females. For male mice, 3% of the control and 24% of the rapamycin-treated animals were alive at the age of 90% mortality. Practically the same survival rate was in females. The causes of death were similar in control and rapamycin-treated mice.

In our study [20], fifty-eight 2-month-old female FVB/N transgenic HER-2/neu mice were randomly divided into two groups. The first group of animals received 1.5 mg/kg rapamycin subcutaneously (s.c.) 3 times a week for a period of 2 weeks followed by 2-week intervals. In the second group, mice received solvent without rapamycin and served as controls. Treatment with rapamycin significantly inhibited age-related body weight gain. While control mice constantly gained weight during their life span, mice that received rapamycin demonstrated a very modest weight increase. Most importantly, in the control group, only 4 mice survived until the age of 11 months (14.3%) compared to 13 (43.3%) in the rapamycin-treated group (p < 0.001). Rapamycin treatment slightly increased the mean (+4.1%) and maximal life span (+12.4%). The mean life span of long-living animals (last 10% of survivors) was significantly greater in the group receiving rapamycin (+11%) compared to control. Parameter α of the Gompertz model, which is interpreted as the rate of demographic aging, was 1.8 times lower in the group subjected to rapamycin treatment than in control. A half of control mice develop mammary adenocarcinomas (MAC) by day 206, whereas in rapamycin-treated group, this period was extended to 240 days. Remarkably, rapamycin decreased the mean number of tumors per tumor-bearing mouse by 33.7% and the mean size of MAC by 23.5%.

In another our study, 66 female 129/Sv mice at the age of 2 months were randomly divided into two groups. The first group of animals received rapamycin as HER-2/neu mice, and the second group served as controls. Treatment with rapamycin significantly inhibited age-related weight gain in female mice [21]. While control mice constantly gained weight during their life span, mice that received rapamycin demonstrated a very modest weight increase. As a result, from the 5th to the 23rd month, body weight was increased by 21.9% and 12.4% in the control and rapamycin-treated groups, respectively. The body weight in the rapamycin-treated animals was significantly less compared with the control between the age of 20 and 27 months. Rapamycin slightly affected food consumption in young mice and decreased food consumption by 23% in old mice. There was no significant difference in age-related dynamics of the length of the estrous cycle and in the ratio between the estrous cycle phases in the control and rapamycin-treated groups. However, at the age of 18 months, 46 and 65% of mice had a regular estrous cycle in the control and rapamycin-treated groups, respectively. Most importantly, 35.5% of control mice survived until the age of 800 days compared to 54.3% in the rapamycin-treated group, whereas until the age of 900 days – 9.7 and 31.4%, respectively (p < 0.01). Twenty-tree percent of female mice exposed to rapamycin survived the age at death of the last mouse in the control group. Rapamycin significantly decreased the incidence of spontaneous tumors in these mice as well.

In the study of Komarova et al. [22], rapamycin was given in drinking water to 35 male mice heterozygous for a germline p53 null allele ($p53^{+/-}$) beginning at various ages. Thirty-eight intact mice served as controls. The mean life span of animals in the control group was 373 days, whereas in rapamycin-treated mice 410 days. Spontaneous carcinogenesis was significantly delayed in rapamycin-treated mice compared to control mice. Then, during analysis of the results, all mice were subdivided into two groups: 'young' (receiving rapamycin from the age of 5 months or earlier) and 'old' (receiving rapamycin starting at 5 months of age or older). The mean life span in rapamycin-treated 'young' mice reached 480 days, a 3.5-month increase over the control group. Thus, the life-extending effect of rapamycin is more pronounced in the group exposed to the treatment earlier in life.

Nanoformulated micelles of rapamycin, Rapatar, were given as gavage to p53 null ($p53^{-/-}$) male mice [23]. The treatment with Rapatar extended the mean life span by 30% and delayed tumor development in highly tumor-prone $p53^{-/-}$ mice. Mean tumor latencies for the control $p53^{-/-}$ and Rapatar-treated $p53^{-/-}$ mice were 161 and 261 days, respectively.

Beginning at 9 weeks of age until natural death, $Rb1^{+/-}$ (B6.129S2(Cg)-Rb1^{tm1Tyj}) and $Rb1^{+/+}$ mice were fed a diet without or with enterically released formulation of rapamycin [24]. The mean life span of rapamycin-treated female $Rb1^{+/-}$ mice was increased by 8.9% and by 13.8% in males as compared with the sex-matched controls. Once all $Rb1^{+/-}$ mice had died, the $Rb1^{+/+}$ littermates were euthanized. Approximately 85% of rapamycin-treated versus 50% of controls survived this age in females, whereas 60 versus 25%, respectively, in males. Thus, the life span was extended more in female than in male wild-type littermates. Exposure to rapamycin was followed by inhibition of C-cell thyroid carcinomas in $Rb1^{+/-}$ mice.

Conclusion

Data on physiological and molecular mechanisms of the beneficial effects of biguanides and rapamycin on life span and inhibitory tumorigenesis have been discussed in several recent papers [3, 10, 17, 22] and are summarized in table 1. The effects of biguanides and rapamycin are presented in accordance with their levels of integration: molecular, cellular/tissue and systemic/organismal. There is a significant similarity in the majority of effects of these two groups of drugs and in the main patterns of their activities as antiaging and anticarcinogenic remedies. It means that the key targets as well as signaling pathways and regulatory signals are also similar. Moreover, there is also sufficient similarity in patterns of changes observed during normal aging and in the process of carcinogenesis. DNA damage response signaling seems to be a key mechanism in the establishment and maintenance of senescence as well as carcinogenesis. Some aspects of the problem have been discussed elsewhere [2]. The available data on cellular senescence in vitro and on accumulation of various human pre-malignant lesions in the cells in vivo provide evidence suggesting that senescence is an effective natural cancer-suppressing mechanism [25]. At the same time, adequate clinical application of therapy-induced 'accelerated senescence' for prevention, progression, or recurrence of human cancers is still insufficiently understood. The mechanisms underlying the bypass of senescence response in the progression of tumors still have to be discovered. Recent studies reveal a negative side of cellular senescence, which is associated with the secreted inflammatory factors, and may alter the microenvironment in favor of cancer progression designated as syndrome of cancerophilia [26] or senescence-associated secretory phenotype (SASP) [4]. Thus, cellular senescence suppresses the initiation stage of carcinogenesis, but is the promoter for initiated cells. We believe that the similarity between two fundamental processes – aging and carcinogenesis – is a basis for understanding the two-side effects of biguanides and rapamycin (fig. 1). The reasons for the difference in response to metformin and rapamycin in different strains of mice are not well understood. Recent findings provide evidence for inhibitory effects of metformin and rapamycin on the SASP interfering with IKK-β/NF-κB [4, 26] – an important step in the hypothalamic programming of systemic aging [27] and in carcinogenesis [2]. It remains to be shown whether antidiabetic biguanides and rapamycin can extend life span in humans.

Acknowledgements

This paper was supported in part by a grant 6538.212.4 from the President of the Russian Federation. The author is very thankful to Dr. Tatiana V. Pospelova and Dr. Mark A. Zabezhinski for critical reading of the manuscript and valuable comments.

References

1 López-Otín C, Blasco MA, Partridge L, Serrano M, Kroemer G: The hallmarks of aging. Cell 2013;152: 1194–1217.

2 Anisimov VN: Carcinogenesis and aging 20 years after: escaping horizon. Mech Ageing Dev 2009;130: 105–121.

3 Anisimov VN, Bartke A: The key role of growth hormone-insulin-IGF-1 signaling in aging and cancer. Crit Rev Oncol Hematol 2013;87:201–223.

4 Johnson SC, Rabinovitch PS, Kaeberlein M: mTOR is a key modulator of ageing and age-related disease. Nature 2013;493:338–345.

5 Selman C, Tullet JM, Wieser D, Irvine E, et al: Ribosomal protein S6 kinase 1 signaling regulates mammalian life span. Science 2009;326:140–144.

6 Lamming DW, Ye L, Katajisto P, Goncalves MD, Saitoh M, Stevens DM, Davis JG, Salmon AB, Richardson A, Ahima RS, Guertin DA, Sabatini DM, Baur JA: Rapamycin-induced insulin resistance is mediated by mTORC2 loss and uncoupled from longevity. Science 2012;335:1638–1643.

7 Spindler SR: Caloric restriction: from soup to nuts. Ageing Res Rev 2010;9:324–353.

8 Mattison JA, Roth GS, Beasley TM, Tilmont EM, Handy AM, Herbert RL, Longo DL, Allison DB, Young JE, Bryant M, Barnard D, Ward WF, Qi W, Ingram DK, de Cabo R: Impact of caloric restriction on health and survival in rhesus monkeys from the NIA study. Nature 2012;489:318–321.

9 Anisimov VN, Semenchenko AV, Yashin AI: Insulin and longevity: antidiabetic biguanides as geroprotectors. Biogerontology 2003;4:297–307.

10 Anisimov VN, Berstein LM, Egormin PA, Piskunova TS, Popovich IG, Zabezhinski MA, Kovalenko IG, Poroshina TE, Semenchenko AV, Provinciali M, Re F, Franceschi C: Effect of metformin on life span and on the development of spontaneous mammary tumors in HER-2/neu transgenic mice. Exp Gerontol 2005;40:685–693.

11 Anisimov VN, Egormin PA, Piskunova TS, Popovic IG, Tyndyk ML, Yurova MV, Zabezhinski MA, Anikin IV, Karkach AS, Romanyukha AA: Metformin extends life span of HER-2/neu transgenic mice and in combination with melatonin inhibits growth of transplantable tumors in vivo. Cell Cycle 2010;9: 188–197.

12 Anisimov VN, Berstein LM, Egormin PA, Piskunova TS, Popovich IG, Zabezhinski MA, Tyndyk ML, Yurova MV, Kovalenko IG, Poroshina TE, Semenchenko AV: Metformin slows down aging and extends life span of female SHR mice. Cell Cycle 2008;7:2769–2773.

13 Anisimov VN, Berstein LM, Popovich IG, Zabezhinski MA, Tyndyk ML, Egormin PA, Piskunova TS, Yurova MN, Semenchenko AV, Kovalenko IG, Poroshina TE: If started early in life, metformin increases life span and postpones tumors in female SHR mice. Aging (Albany) 2011;3:148–157.

14 Anisimov VN, Piskunova TS, Popovich IG, Zabezhinski MA, Tyndyk ML, Egormin PA, Yurova MN, Semenchenko AV, Kovalenko IG, Poroshina TE, Berstein LM: Gender differences in metformin effect on aging, life span and spontaneous tumorigenesis in 129/Sv mice. Aging (Albany) 2010;2:945–958.

15 Ma TC, Buescher JL, Oatis B, Funk JA, Nash AJ, Carrier RL, Hoyt KR: Metformin therapy in a transgenic mouse model of Hungtington's disease. Neurosci Lett 2007;411:98–103.

16 Smith DL, Elam CF, Mattison JA, Lane MA, Roth GS, Ingram DK, Allison DB: Metformin supplementation and life span in Fischer-344 rats. J Gerontol Biol Sci 2010;65A:468–474.

17 Martin-Montalvo A, Mercken EM, Mitchell SJ, Palacios HH, et al: Metformin improves healthspan and lifespan in mice. Nat Commun 2013;4:2192.

18 Harrison DE, Strong R, Sharp ZD, Nelson JF, Astle CM, Flurkey K, Nadon NL, Wilkinson JE, Frenkel K, Carter CS, Pahor M, Javors MA, Fernandez E, Miller RA: Rapamycin fed late in life extends lifespan in genetically heterogeneous mice. Nature 2009;460:392–396.

19 Miller RA, Harrison DE, Astle CM, Baur JA, Boyd AR, de Cabo R, Fernandez E, Flurkey K, Javors MA, Nelson JF, Orihuela CJ, Pletcher S, Sharp ZD, Sinclair D, Starnes JW, Wilkinson JE, Nadon NL, Strong R: Rapamycin, but not resveratrol or simvastatin extend life span in genetically heterogeneous mice. J Gerontol A Biol Sci Med Sci 2011;66:191–201.

20 Anisimov VN, Zabezhinski MA, Popovich IG, Piskunova TS, Semenchenko AV, Tyndyk ML, Yurova MN, Antoch MP, Blagosklonny MV: Rapamycin extends maximal life span in cancer-prone mice. Am J Pathol 2010;176:1092–2096.

21 Anisimov VN, Zabezhinski MA, Popovich IG, Piskunova TS, Yurova MN, Semenchenko AV, Tyndyk ML, Yurova MN, Rosenfeld SV, Blagosklonny MV: Rapamycin increases lifespan and inhibits spontaneous tumorigenesis in inbred female mice. Cell Cycle 2011;10:4230–4236.

22 Komarova EA, Antoch MP, Novototskaya LR, Chernova OB, Paszkiewicz G, Leontieva OV, Blagosklonny MV, Gudkov AV: Rapamycin extends lifespan and delays tumorigenesis in heterozygous $p53^{+/-}$ mice. Aging (Albany) 2012;4:709–717.

23 Comas M, Toshkov I, Kuropatwinski KK, Chernova OB, Polinsky A, Blagosklonny MV, Gudkov AV, Antoch MP: New nanoformulation of rapamycin Rapatar extends lifespan in homozygous *p53*^{-/-} mice by delaying. Aging (Albany) 2012;4:715–722.

24 Livi CB, Hardman RL, Christy BA, Dodds SG, Jones D, Williams C, Strong R, Bokov A, Javors MA, Ikeno Y, Hubbard G, Hasty P, Sharp ZD: Rapamycin extends life span of *Rb1*^{+/-} mice by inhibiting neuroendocrine tumors. Aging (Albany) 2013;5:100–110.

25 Demidenko ZN, Zubova SG, Bukreeva EI, Pospelov VA, Pospelova TV, Blagosklonny MV: Rapamycin decelerates cellular senescence. Cell Cycle 2009;8: 1888–1895.

26 Dilman VM: Development, Aging and Disease. A New Rationale for an Intervention. Chur, Harwood Academic Publishers, 1994.

27 Zhang G, Li J, Purkayastha S, Tang Y, Zhang H, Yin Y, Li B, Liu G, Cai D: Hypothalamic programming of systemic ageing involving IKK-β, NF-κB and GnRH. Nature 2013;497:211–216.

Vladimir N. Anisimov
Department of Carcinogenesis and Oncogerontology
N.N. Petrov Research Institute of Oncology, 68, Leningradskaya St.
Pesochny-2, St. Petersburg, 197758 (Russia)
E-Mail aging@mail.ru

Author Index

Subject Index

Network analysis, *see specific networks*; Systems biology
Neuroendocrine Theory of Aging 58, 59
Nonalcoholic fatty liver disease (NAFLD), metabolic syndrome 103, 104
Nuclear factor-κB (NF-κB), inflammation role 102, 104

Obesity
    inflammation role 101, 102
    mammalian target of rapamycin role 121, 122
Ordinary differential equations (ODEs), computational systems modeling 39
Oxidative stress
    cell damage and aging 133, 134
    epigenetic clock hypothesis 59
    melatonin free radical scavenging 137
    reactive oxygen species dose 165, 166
    ROSage project 166–174

p21 171
p53
    Alzheimer's disease studies 40
    knockout mouse 114, 185
Parkinson's disease (PD) 36, 115, 121
Partial differential equations (PDEs), computational systems modeling 39, 40
Pascal's Wager 50
Peroxynitrite, cell damage and aging 130–132
Petri net 41
Phenformin, geroprotection 180–182
Phosphatidylinositol-3-kinase (PI3K) 43, 44, 179
Power plot, scale-free networks 10, 11, 25
PRAS40 44
Prebiotics, gut microbiota in the elderly effects 150–152

Random network 9
Rapamycin, aging and longevity effects 184–186
Raptor 108, 180
Reactive oxygen species, *see* Oxidative stress
Regular network 9
Reliability theory, *see* Systems biology
Replicative life span (RLS) network, *Saccharomyces cerevisiae* 19–21, 24, 28, 29
Retinoblastoma protein (Rb), knockout mouse 185, 186
Rheb 111

Ribosome capacity, mTORC signaling 110
Rictor 108, 180
ROSage project 166–174

S6 kinase 114, 122, 179, 180
Scale-free network 9–11, 24, 25
Senescence-associated secretory phenotype (SASP) 130, 133, 186
SENS hypothesis, aging 162
SIRT1 129, 134, 136
Small-world network 9, 11, 25–27
SREBP 122
Stem cell, *see also* Hematopoietic stem cell
    histone modification and stem cell fate 164, 165
    mammalian target of rapamycin in maintenance 117
Suprachiasmatic nucleus (SCN), circadian clock regulation 129
Systems biology, *see also* Computational systems modeling; *specific models and networks*
    advantages in aging research 37, 38
    aging as systemic phenomenon 161, 162
    clusters and hierarchies 12, 76
    complex systems
        aging phenotypes 159, 160
        nonlinear network complexity 158
        properties 3, 4
    complicated versus complex systems 3
    conceptual workflows 155–157
    evolvability of system 14
    frailty 12, 13, 86–89
    graphs
        components 8, 9, 19
        network structure and connectivity 9, 10, 19
        node-node connectivity 11, 12
        power plots and scale-free networks 10, 11, 25
        small-world network categorization 11, 25–27
    large network reconstruction and simulation 158
    multi-scale data integration and modeling 158
    nonlinear dynamics in aging networks 160, 161
    nonlinear systems theory 4, 5
    omics approach 157, 158
    overview 5, 6